INTRODUCTION TO PROGRAMMABLE LOGIC CONTROLLERS

Third Edition

Programming the SLC 500 PLC
Using
RSLogix 500 Software

Lab Manual

Gary Dunning

DELMAR
CENGAGE Learning™

Australia • Brazil • Japan • Korea • Mexico • Singapore • Spain • United Kingdom • United States

Introduction to Programmable Logic Controllers: Lab Manual, Third Edition
Gary Dunning

Vice President, Technology and Trades SBU:
 Alar Elken

Editorial Director: Sandy Clark

Senior Acquisitions Editor: Stephen Helba

Developmental Editor: Sharon Chambliss

Marketing Director: David Garza

Channel Manager: Dennis Williams

Marketing Coordinator: Stacey Wiktorek

Production Director: Mary Ellen Black

Production Editor: Barbara L. Diaz

Editorial Assistant: Dawn Daugherty

Library of Congress Control Number: 2005028293

ISBN-13: 978-1-4018-8427-7

ISBN-10: 1-4018-8427-X

Delmar
Executive Woods
5 Maxwell Drive
Clifton Park, NY 12065
USA

Cengage Learning is a leading provider of customized learning solutions with office locations around the globe, including Singapore, the United Kingdom, Australia, Mexico, Brazil, and Japan. Locate your local office at **www.cengage.com/global**

Cengage Learning products are represented in Canada by Nelson Education, Ltd.

To learn more about Delmar, visit **www.cengage.com/delmar**

Purchase any of our products at your local bookstore or at our preferred online store **www.ichapters.com**

Printed in the United States of America
3 4 5 6 7   13 12 11 10 09
ED357

TABLE OF CONTENTS

PREFACE

Lab exercises in this manual assume you have probably not programmed an SLC 500 or MicroLogix PLC using a personal computer and software before. Since this is an introduction to programming a programmable logic controller, this manual will show you how to program basic ladder diagrams. We will not cover all PLC features, instructions, or system configurations. The purpose of these exercises is to provide your first hands-on programming exercises and PLC interface experience.

As of this writing, there are five SLC 500 modular processors: the fixed SLC 500, and the MicroLogix 1000, 1200, and 1500. Your personal computer operating system can be Windows 95, Windows 98, Windows NT, Windows 2000, or Windows XP. Your computer can be a desktop, a laptop, a notebook, or possibly an industrial computer. There are also many ways to configure your personal computer to connect to and communicate with the particular PLC you are working with. We will work with a personal computer and PLC interface in Lab Exercise 5.

Keep in mind that there are many different versions of RSLogix 500 software. As of this writing, the current version of RSLogix 500 is version 6.30. Because many people do not have the latest software, we have left the lab manual at version 4.50. If you have version 4.50 or later, you will be able to complete all the labs—even though some of the screens pictured in the lab manual may not look exactly like those in your particular version of RSLogix 500 software. Procedures for navigating to specific screens may also have changed. Ask your instructor for assistance.

This lab manual starts with the following assumptions:

1. Your instructor has loaded Rockwell Software's RSLogix 500 software on your personal computer. We will be using RSLogix version 4.50 software.
2. Your instructor has demonstrated which method of personal computer-to-PLC interface is preferred in your lab setup.
3. Your SLC 500 hardware consists of an SLC 5/01, 5/02, 5/03, 5/04, or 5/05 modular processor in any of the available chassis, or a MicroLogix 1000 PLC.

HOW TO DO THE LAB EXERCISES

Each lab exercise will be divided into the following sections:

1. Objectives
2. Introduction
3. Equipment required, if additional materials are needed
4. The lab(s)
5. Review questions

CONVENTIONS

Introductory lab exercises will lead you through each key closure or mouse click needed to develop the program. You will enter most information by using your mouse or keyboard keys. Each step will have a number along with a check-off area. The check-off area is to help you track which steps you have completed. The numbered steps below represent a sample of the steps you may execute as part of the "The Lab" section.

1. _____ Start from the main screen from your RSLogix 500 software.
2. _____ Insert your floppy disk in drive A.
3. _____ Click on the Open File icon.

4. _____ From the Open/Import SLC 500 Program window, click on the down arrow on the right side of the look-in/drop-down list box.

5. _____ Select Floppy A.

6. _____ Click on the Begin file.

7. _____ Click Open.

In most cases, after you have entered the introductory program, you will be asked to develop a similar program on your own. There will be review questions at the end of selected lab exercises.

1

Introduction to SLC 500 Hardware

OBJECTIVES

Upon completion of this laboratory exercise, you should be able to:

- identify SLC 500 PLC hardware
- identify your SLC 500 modular processor or MicroLogix fixed PLC

If you are using a modular SLC 500 PLC, you should be able to:

- identify a modular chassis
- determine which modular power supply is attached to your chassis
- identify the I/O modules residing in your SLC 500 chassis
- determine which modular processor resides in the SLC 500 chassis

INTRODUCTION

This exercise will familiarize you with the SLC 500 or MicroLogix hardware you will be using while working through the lab exercises.

You will be using one of the following members of the SLC 500 family: the MicroLogix 1000, the fixed I/O PLC, or the SLC 500 modular PLC with a 5/01, 5/02, 5/03, 5/04, or 5/05 processor.

MODULAR SLC 500 PROCESSOR CONNECTIVITY

5/05 Processor	Ethernet and a serial port
5/04 Processor	Data Highway Plus and a serial port
5/03 Processor	Data Highway 485 and a serial port
5/02 Processor	Data Highway 485 port only
5/01 Processor	Data Highway 485 port only

SLC FAMILY HARDWARE

A programmable logic controller, usually called a PLC or more commonly a programmable controller, is a solid-state digital industrial computer. Figure 1-1 illustrates the family of Allen-Bradley MicroLogix 1000 PLCs. The top unit is a 20-input and 12-output MicroLogix 1000 microcontroller. The center two MicroLogix 1000 microcontrollers are 10-input and 6-output units. A handheld programming terminal is pictured at the bottom.

Figure 1-1 MicroLogix 1000 fixed I/O PLC. (Used with permission of Rockwell Automation, Inc.)

Modular I/O PLCs

A modular PLC does not have the I/O terminal strips built into the processor unit. Modular PLCs have their I/O points on plug-in type removable units called I/O modules. SLC 500 PLCs with modular inputs and outputs consist of a chassis and a modular power supply. A selection of five modular CPUs and all input and output modules are separate hardware items. Figure 1-2 shows a modular Allen-Bradley SLC 500 PLC, with its parts labeled, assembled together to make a working PLC.

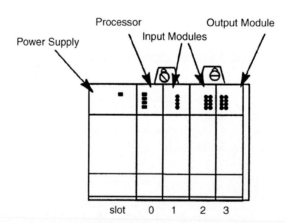

Figure 1-2 Allen-Bradley SLC 500 modular PLC with major pieces identified. (Used with permission of Rockwell Automation, Inc.)

When I/O is modular, the user can mix input and output types in the chassis, to meet specific needs. There are usually few limitations on the mix or positioning of I/O modules. Common SLC 500 chassis consist of 4, 7, 10, or 13 slots. One slot is for the processor and the remaining slots are for I/O modules. Typical modules will contain 4, 8, 16, or 32 I/O points.

Electrical connections between each module and the CPU are made by two mating plugs. One plug is located on a printed circuit board sticking out the back of each module. A printed circuit board that the modules plug into runs along the back of the chassis and is called the backplane. Refer to your text, Figures 1-24 and 1-25.

Default SLC 500 Hardware Used in This Manual

The default SLC 500 in a seven-slot chassis used to develop the lab exercises for this lab manual is listed in the table in Figure 1-3. A MicroLogix 1000 PLC with analog capabilities can also be used for the labs.

DEFAULT SLC 500 PLC USED TO DEVELOP LAB EXERCISES		
Slot	Module Part Number	I/O Points
0	Processor SLC 5/04	None
1	1746-IB16	16 Inputs
2	1746-OB16	16 Outputs
3	1746-OB16	16 Outputs
4	1746-IB8	8 Inputs
5	1746-OB16	16 Outputs
6	1746-NIO4V	2 Channel analog inputs (−20 mA to +20 mA or +/− 10 Vdc) 2 Channel analog outputs (+/− 10 Vdc)

Figure 1-3   Default SLC 500 PLC.

INSTRUCTOR'S DEMONSTRATION

Follow along while your instructor demonstrates and passes around SLC 500 hardware.

THE LAB

1. _____ Go to your lab station and identify your PLC hardware.
2. _____ Locate the SLC 500 processor or MicroLogix and list it here. _____
_____
3. _____ Determine the inputs and outputs.
4. _____ Identify how your particular field devices wire to your input and output screw terminals.

If you will be using a modular SLC PLC, be sure you can identify the following elements.

5. _____ The chassis in your lab demo unit. How many slots are in your SLC 500 chassis: 4, 7, 10, or 13? Circle the correct answer.
6. _____ As you look at the chassis, the power supply is on the left side. Identify the power supply part number. List that number here: _____
7. _____ Each position in the chassis that will accept a module is called a slot. The first slot to the right of the power supply is slot zero. Refer to Figure 1-4. The SLC 500 modular processor always resides in slot zero. Identify the modular SLC 500 processor and list it in slot zero of the table shown in Figure 1-5.

Power Supply	Slot 0	Slot 1	Slot 2	Slot 3	Slot 4	Slot 5	Slot 6
	CPU	I/O	I/O	I/O	I/O	I/O	I/O

Figure 1-4   SLC 500 modular chassis power supply placement along with chassis slot identification for a seven-slot chassis.

8. _____ The first module to the right of the processor starts as slot one and increments to the right. Identify I/O modules and list them in Figure 1-5.

THE SLC 500 PLC I WILL USE TO COMPLETE LAB EXERCISES		
Slot	**Module Part Number**	**I/O Points**
0		None
1		
2		
3		
4		
5		
6		
7		
8		
9		

Figure 1-5 My SLC 500 PLC.

Some specifications of the SLC 500 PLC you will be using to complete these lab exercises are now listed in Figure 1-5.

9. _____ Identify input and output screw terminals.
10. _____ Identify removable I/O terminal blocks.
11. _____ Identify the communication ports on your modular processor.
12. _____ Where is the communication cable attached to your processor?
13. _____ Determine where the communication cable is connected to your personal computer.

For this lab you need only determine where the cables are connected. We will look at communications in depth in a later lesson.

2

Introduction to Windows and the RSLogix 500 Software

OBJECTIVES

Upon completion of this laboratory exercise, you should be able to:

- open the RSLogix 500 software and navigate
- use the help screens to assist in answering your software questions
- navigate in the project and ladder windows
- set up ladder window parameters
- open different data or ladder files and display their properties

INTRODUCTION

Welcome to RSLogix 500 software. This lab manual is intended as an introduction to programming, editing, and maintaining MicroLogix, fixed SLC 500, and modular SLC 500 PLC programs. RSLogix 500 software is a 32-bit Windows programming package used to develop ladder logic programs for the SLC 500 family and the MicroLogix family of PLCs. To help you get started, we will review the basics of navigating through the Windows environment using a mouse. We will also introduce the various Windows and toolbars in the RSLogix 500 software. It is important that you understand what the various windows and toolbars contain before you begin developing SLC 500 projects. An SLC 500 project is the collection of program files, processor properties, and data files created for a ladder program. The project contains all ladder instructions, data, and configuration information associated with your user program.

NEW TERMS

Program Files	Program files contain the main ladder program and subprograms called subroutine program files.
Data Files	Data files contain data that is used in conjunction with ladder program instructions. Examples of data files include input and output files, and timer, counter, and integer files.
Project	An SLC 500 project is the collection of program files, processor properties, and data files created for a ladder program. The project contains all ladder instructions, data, and configuration information associated with your user program.

5

Database	Text associated with your ladder program is stored in the project database. All database information is stored on your personal computer's hard drive only. Database information is never transferred to the SLC 500 family processor. Database information is comprised of: addresses and symbols, which include I/O addresses and text symbols associated with the I/O address instruction comments; instruction comments, which include text associated with a specific instruction and address pair or address instruction pair; and rung comments and page titles. A rung comment is text associated with a ladder rung. Page titles provide the opportunity to separate ladder logic into groups of rungs called pages.
Processor File	The processor file is the project information that can be transferred from your personal computer to the PLC processor. This file includes all project information except the database files.
Configuration Information	The programming software needs to know what PLC hardware is being used in conjunction with the ladder program. Configuration information includes information about the processor, I/O module types, communication configuration information, and so on.

EQUIPMENT REQUIRED

For this exercise you will need a personal computer with a recent version of Rockwell RSLogix 500 software.

Personal Computer Hardware

The minimum computer system requirements for installing and running RSLogix 500 software on a personal or industrial computer are as follows:

- starting with RSLogix version 5.50, Windows 95 is no longer supported
- Windows 98, ME, NT 4.0 SP 6a, 2000, or XP
- personal computer with Pentium 100 M hz or higher
- minimum 32-MB RAM
- hard disk with minimum 35-MB free space
- 3.5-inch high density (1.44 MB) disk drive and /or CD-ROM drive
- monitor supported by Windows, SVGA 16 color, 800 × 600

To effectively use the software, you should be familiar with Windows operations. You should also know how to use a mouse, choose menu commands, and work with Windows and dialog boxes.

To execute many commands you will use a mouse to point to a button or menu command and then click. When using a mouse, you should hold the mouse perpendicular to your monitor so that moving the mouse sideways also moves the mouse pointer sideways on the screen. When the mouse pointer is positioned over the desired object, you will click either the right or left mouse button. Once in position, you click once to select an item, and double-click to execute an action or command. The table in Figure 2-1 lists and defines the basic mouse commands.

Windows Dialog Boxes

Windows dialog boxes are used with Windows commands when additional information is required before a command, such as printing or opening a program, can be completed. Not all dialog boxes have all the features identified in the following paragraphs. Familiarize yourself with

BASIC MOUSE COMMANDS	
Term	**Definition**
Click	Position the mouse pointer on an object, area, or field, then press and release the mouse button once.
Double-click	Position the mouse pointer on an object, area, or field, then press and release the mouse button twice quickly.
Right-click	Position the mouse pointer on an object, area, or field, then press and release the right mouse button once.
Left-click	Position the mouse pointer on an object, area, or field, then press and release the left mouse button once.
Select item or command	Click to highlight the item to be affected by the next command, or click on a dialog box option.
Choose item or command	Click on a tool, menu command, or an item in a dialog box.
Drag and drop	Dragging and dropping is a four step process that moves an object from one point on your screen to another. 1. Point to the object. 2. Press and hold the left mouse button. 3. Move the mouse pointer to the location to which you'd like to move the selected object. 4. Release the mouse button to drop the object into its new location.

Figure 2-1   Basic mouse commands.

the dialog box features in Figure 2-2. A sample screen printout of a Windows dialog box from RSLogix 500 software is shown.

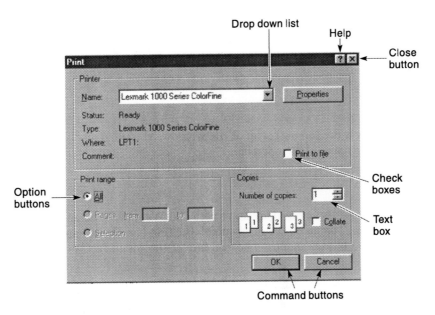

Figure 2-2   Windows dialog box.

RSLogix 500 Main Window

Each time you create a new project or open an existing RSLogix project the main window opens; this is where you will create your PLC ladder project. Figure 2-3 illustrates RSLogix' main window. Each window feature is identified on pages 8 through 12.

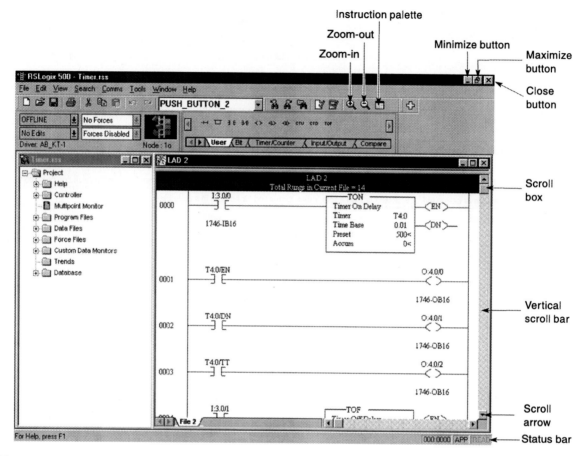

Figure 2-3 RSLogix 500 software main window with features identified.

Window Title Bar

The title bar is the topmost strip of the window. The title bar displays the name of the program as well as the name of the open file.

Menu Bar

The buttons on the menu bar, along with the Windows toolbar, execute commands when you click on them. The menu bar lies beneath the title bar. The menu bar contains key words associated with the hidden menus. Click on any of the menu bar key words to open the associated menu.

Windows Toolbar

The Windows toolbar buttons execute standard Windows commands when you click on them. Pointing to each of the buttons will display their names in little boxes below the button. These are standard Windows buttons that are used to open, save, print, cut, copy and paste, and undo and redo. The window in the center of the Windows toolbar contains information from previous searches of the ladder logic. The next three buttons are used in conjunction with the search window. These buttons are used to find the previous occurrence of the selected search object, the next occurrence of the selected search object, or all occurrences of the selected search object.

The next two buttons (the ones with the check marks) are used to check your ladder program for errors. This is called verifying. The file with the check mark is the verify file button, and the computer icon with the check mark on it will verify the entire project to check for errors. Your project must be verified and all errors corrected before the project can be downloaded and run in the PLC's CPU. The left-hand button is used to verify the program file you are currently working in. The right-hand button is used to verify your entire project.

The buttons with the magnifying glasses can be used to increase or decrease the size of your ladder program rungs. These are the zoom-in or zoom-out buttons.

Program/Processor Status Tool Bar

This toolbar, shown in Figure 2-4, contains four drop-down lists that identify the current processor operating mode. It also identifies current on-line edit status and whether forces are present and enabled, along with the current processor RSLinx driver name and node address.

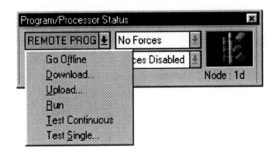

Figure 2-4   Processor status drop-down menu.

Project Window

The project window displays the file folders listed in the project tree.

Project Tree

The project tree is a visual representation of all folders and their associated files contained in the current project.

Results Window

This window displays the results of either a search or a verify operation. The verify operation is used to check the ladder program for errors. Figure 2-5 shows the verify tab indicating a program error.

Notice the Verify Results tab and Search Results tab. Clicking on one or the other will display the contents of the tab. The results window will open automatically whenever there are one or more program errors as a result of verifying a ladder program ladder file or the entire project, or when a "find all" search operation is executed.

Active Tab

The active tab identifies which program file is currently active.

Status Bar

The status bar contains information relevant to the current file. The left side of the task bar contains messages relating to your program. The right side of the task bar contains four informational areas: XREF, 2.0001, APP, and READ, described on the next page.

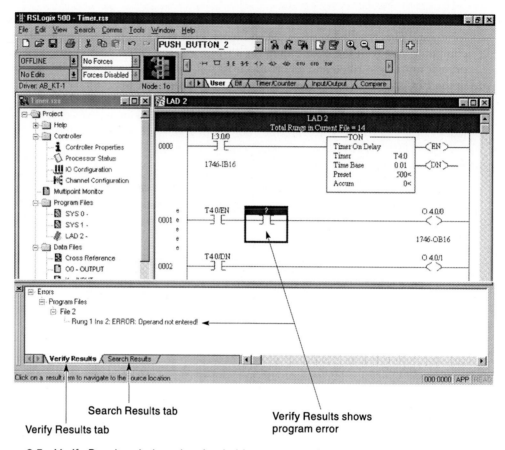

Search Results tab

Verify Results tab

Verify Results shows
program error

Figure 2-5   Verify Results window showing ladder programming error.

1. The XREF section alerts you that current cross-reference information displayed on your ladder needs to be refreshed before it will display correctly. This adjustment is necessary after editing your ladder program.
2. 2.0001 indicates the current cursor position within the ladder program. The format is 0002.0000; the number to the left of the point is the file number, while the number to the right of the point is the current rung number.
3. APP will display as either APP for append or INS for insert. This signifies the current instruction entry mode. Append places instructions after the current instruction on your ladder, while insert places instructions in front of the current instruction.
4. READ, even though normally grayed out, will be displayed black when the ladder program file is either designated as read-only or someone else opened the program file before you and is currently working on this same program file.

Split Bar

The split bar is used to split the ladder window so as to display two different program files, or two different or separated groups of ladder rungs. Figure 2-6 illustrates two ladder program files being displayed after using the Windows split window feature.

Minimize and Maximize Buttons

The minimize button will cause its associated window to disappear and then reappear as a button on the task bar along the bottom of the screen. This does not close the window. The maximize button, on the other hand, causes the current window to enlarge to fill the full screen. Once

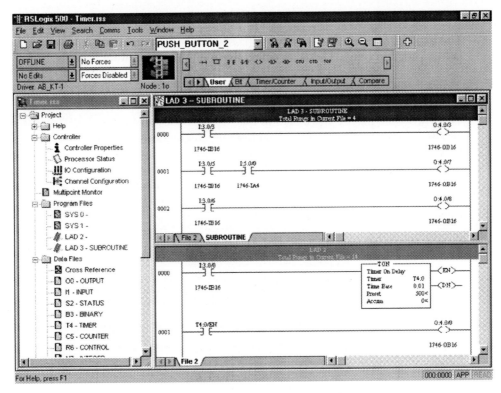

Figure 2-6 Two ladder files displayed in the ladder window.

maximized, the button changes to a restore button. The restore button returns the window to the size it was before it was maximized.

Close Button

The close button removes the window from the computer's memory. To use a closed window, it must be reopened. Reopening a window means it is reloaded into the computer's operating memory (RAM) from the hard drive.

Tabbed Instruction Toolbar

The tabbed instruction toolbar displays the instruction set as a group of tabbed categories. Clicking on a category tab (shown in Figure 2-7) from the tabbed instruction bar changes the instruction toolbar (A), just above it, to display the associated instructions (B). The arrow buttons to the right or left of either the tabbed instruction toolbar or the instruction toolbar allow you to display tabs or instructions for either toolbar that may be hidden (C). Starting with RSLogix software version 4.0, the user tab became customizable. Use the arrows (D) to display hidden instructions.

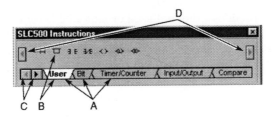

Figure 2-7 RSLogix tabbed instruction toolbar with the user tab selected.

Zoom-in/Zoom-out Buttons

The two buttons with the magnifying glasses can be used to increase or decrease the size of your ladder program rungs. These are called the zoom-in or zoom-out buttons.

Instruction Palette

The instruction palette was added to the RSLogix 500 software with RSLogix version 3. It offers another method of accessing programming instructions. The major feature of the instruction palette is that all the available instructions are displayed in one table. This makes programming easier, as the programmer does not have to move from tab to tab to select instructions from the instruction palette. This instruction palette is also customizable. Instructions that are seldom or never used may be removed from the palette. The instruction palette is illustrated in Figure 2-8.

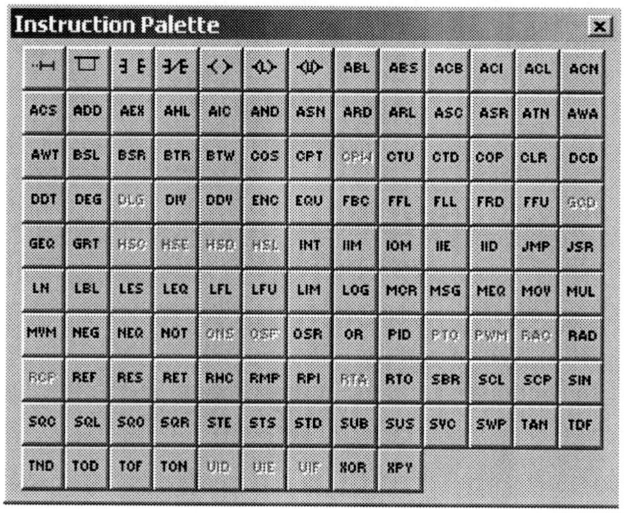

Figure 2-8 RSLogix 500 Instruction Palette.

Some of the instructions may be grayed out on the instruction palette. These instructions are not available because the particular processor selected for the current project does not support certain instructions.

Note

The instruction palette and the user tab of the tabbed instruction toolbar can be customized on version 4.0 (and later versions) of the RSLogix software. Customizing the user tab of the tabbed instruction toolbar allows the programmer to position frequently used instructions on the user tab. This will minimize moving from tab to tab when selecting instructions. Likewise, the instruction customization feature for the instruction palette affords the programmer the flexibility of removing instructions either not used or unavailable for the particular processor in use for the current project. The customization feature will be found under View and Toolbars on the Windows menu bar.

Project Tree

The project tree is a visual representation of all the files for the project named in the project tree title bar. See Figure 2-9. The project tree consists of folders and files that contain all the information about the processor, channel configuration, ladder programs, and data associated with the project. From the project tree you can perform the following operations:

- open files
- create files
- modify file parameters
- copy files
- hide or unhide files
- delete files
- rename files

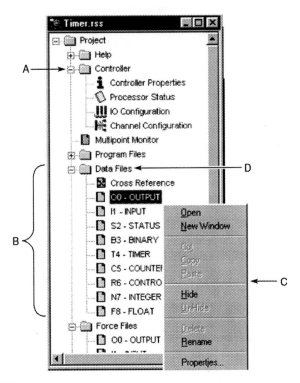

Figure 2-9   RSLogix 500 project tree and sample of a right-click menu.

Notice that in front of each folder there is a plus or minus sign. The plus sign means that the folder is closed. Click on it to open the folder. As an example, (A) in Figure 2-9 identifies the Controller file folder as opened. The contents of the Controller file are displayed. (B) shows the data file folder opened. The data file folder lists all data files associated with this project. (C) shows the output data file right-click options. Likewise, right-clicking on a specific file brings up a menu that allows you to open the file; cut, copy, or paste data; hide or unhide the file; delete or rename the file; and also to view the specific file's properties. The minus sign means that the folder is open and all files are visible. Click on it and the files will disappear. Right-clicking on (D) in Figure 2-9 opens a menu where you can create new data files or view the properties of the Data File folder. Double-click on a file to open it. As an example, we will look into Data File N7, the integer file. An integer file is used to store whole number data. Figure 2-10 illustrates Data File N7 opened.

The Properties window in Figure 2-10 (click on A) displays important data about all the elements contained within the file. From the Properties window you can also create new elements or monitor existing data. The Usage button (click on B) displays which file elements are used and

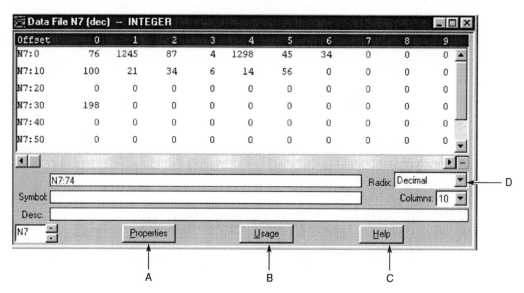

Figure 2-10 RSLogix 500 integer file opened so the data contained can be viewed.

which are available. Clicking on the Help button (C) will display the available help. The radix, or the number system in which data displayed, is determined from the list at (D).

Ladder Window

The ladder window displays the currently open ladder program file. The ladder program is developed and edited from the ladder window. Ladder documentation is also added within this window. Figure 2-11 illustrates a sample ladder window. Notice the ladder rungs and associated text documentation. Right-clicking on any instruction will display a drop-down menu with multiple

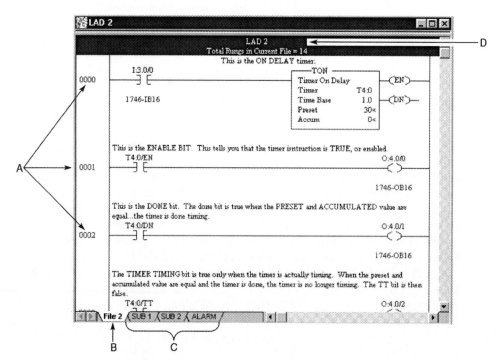

Figure 2-11 RSLogix ladder window.

editing selections. (A) identifies rung numbers. The tab (B) at the bottom identifies this open window as file 2; file 2 is always the main ladder program in the SLC 500 processor. The tabs along the bottom left center (C) represent other subprogram files (subroutines) that are associated with the main ladder program. Clicking on any of the subroutine tabs will display the subroutine ladder rungs. (D) illustrates the page header. The page header identifies which file is open and how many rungs are contained in the ladder file.

Figure 2-11 shows a ladder file window illustrating rungs and their associated documentation along with selected features.

Ladder Window Properties

As the software user, you have the ability to change the way your ladder program and its associated addressing and documentation is displayed. To view the ladder window properties, either right-click on a blank area of the ladder window and select Properties, or select View and then Properties to go to the options window. The ladder options window is displayed in Figure 2-12.

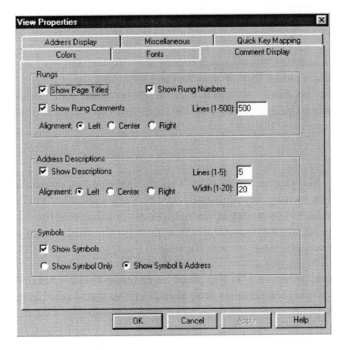

Figure 2-12   RSLogix ladder properties window.

This figure illustrates the six tabs available in the ladder properties options window for RSLogix 500 version 4.5. Depending on the version of your RSLogix 500 software, your tabs may be different. The tabs are described in the following paragraphs.

Comment Display

The Comment Display tab allows you to set up the way program documentation will be displayed on the computer screen. Figure 2-12 illustrates the comment display tab selections. Selections included are:

- will page titles, rung comments, or rung numbers be shown on the screen?
- are address descriptions displayed?
- are symbols to be displayed?

Fonts

The Fonts tab allows selection of the font style and size of text displayed in the ladder window.

Colors

The Colors tab allows you to set up the text and background colors of your software's windows and ladder components. The color displayed for each screen component, such as page titles, descriptions, rung comments, cross-reference information, and symbols, is user configurable.

Address Display

The Address Display tab provides options on how addresses associated with your ladder instructions will be displayed. Will the I/O address be displayed as a single line or as a split line? Will I/O address display include only slot and bit information, or will slot, word, and bit data be displayed? Will bit file addresses be displayed as bit only, or word and bit? Do indirect addresses display on the ladder logic? Will cross-reference information be displayed for inputs or outputs on the ladder rung or not at all?

Miscellaneous

The Miscellaneous tab allows setup of the display. Do you want to display the instructions in three-dimensional format, enable rung wrapping, or display page headers? Will I/O module information be associated with each instruction on your ladder? You can even determine whether, when dragging an instruction from one location on the ladder to another, its address and symbol accompany the dragged instruction.

Quick Key Mapping

Quick Key Mapping is a feature in RSLogix 500 that can streamline your programming tasks by allowing you to map any alphabetic key on your computer keyboard to represent a ladder instruction. Once keys are set up, or mapped, simply depress the keyboard key associated with the desired ladder instruction and it will appear on the rung. As an example, the O key can be assigned (or mapped) to represent the output energize instruction (OTE). The F keyboard key can be mapped to represent the "examine if open" programming instruction (XIO). When programming, simply press the F key to place an examine-if-open instruction on your ladder. Likewise, when the output instruction is programmed, press the O key. Up to twenty-six instructions can be mapped to the twenty-six alpha keys on your computer keyboard.

To enable quick key programming mode, select Edit from the Windows Menu Bar, and select Quick Key Mode. This will insert an empty rung into your program and place the programming software in Quick Key Mode.

INSTRUCTOR'S DEMONSTRATION

Follow along as your instructor demonstrates and navigates through the main features of the RSLogix 500 software.

THE LAB

1. _____ Start your personal computer and run Windows.
2. _____ Click the start button in your Windows software.
3. _____ Click on Programs.
4. _____ Click on Rockwell Software.
5. _____ Click on RSLogix 500 English.

6. _____ Click on RSLogix 500 English. This should launch the RSLogix software.
7. _____ Click on the Open File icon. The Open/Import SLC 500 program dialog box should open as illustrated in Figure 2-13.
    A. List of RSLogix 500 projects.
    B. Click here to select a different drive to find a project.
    C. Demo project to open for this exercise. (Your display may have different files from those displayed.)
    D. Click here to open the selected project.
8. _____ Click on the drop-down list arrow (B) to display the drive and file to look in.
9. _____ Locate the Demo.rss project from the location specified by your instructor.
10. _____ Select the Demo project (C).
11. _____ Click on Open (D).

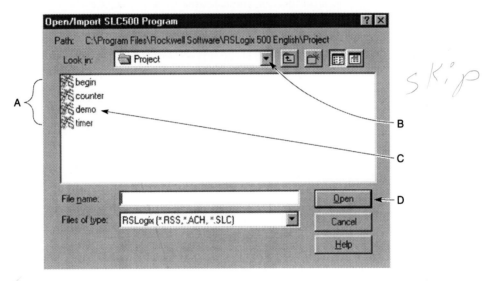

Figure 2-13 The Open/Import SLC 500 program dialog box.

12. _____ Identify the sections of the RSLogix 500 main window referred to in Figure 2-3 on page 8.
13. _____ In the Project Window, use + and − to expand and shrink folders on the Project Tree.
14. _____ Hover over the icons on the Windows toolbar and observe the tool tips displayed, identifying the function of the icon.
15. _____ Click on the Zoom-in and Zoom-out buttons.
16. _____ Scroll through the tabs on the tabbed Instruction toolbar.
17. _____ Click on the Instruction Palette icon to display the Instruction Palette. You may have to resize it when it is displayed on your screen.

Tabbed Instruction Tool Bar

18. _____ On the Tabbed Instruction toolbar, use the right and left arrows to scroll through the instruction tabs.
19. _____ Click on the User tab to display available instructions.
20. _____ Hover your mouse pointer over the normally open contact symbol. Notice the tool tip displays "Examine if Closed." This is the name of this particular instruction.
21. _____ Click on another instruction tab. Observe the available instruction abbreviations; they are called mnemonics.

Ladder Window and Project Window Selection

22. _____ Left-click on a blank spot in the Project Window to make it active.
23. _____ Left-click on a blank spot in the Ladder Window to make it active. In order to work in either of these windows, they must be active. Notice the title bar at the top of the window changes color to identify which window is active.

Ladder File Properties

24. _____ Make the Ladder Window active.
25. _____ Right-click on a blank spot on your ladder window to display the right-click menu. Click on properties.
26. _____ Your display should look similar to Figure 2-12, depending on your version of the RSLogix 500 software.
27. _____ Click on the Comment Display tab.
28. _____ Uncheck Show Page Titles.
29. _____ Click OK to return to your ladder window.
30. _____ Notice that some of the text on your ladder program rungs is gone. These single lines of text are called Page Titles. Go back to the View Properties screen.
31. _____ Recheck Show Page Titles.
32. _____ Uncheck Show Rung Comments and Show Rung Numbers.
33. _____ Click OK to return your ladder window. Notice that the rung comments and rung numbers are now gone.
34. _____ Return to the View Properties screen and recheck the unchecked options.
35. _____ Click on the Colors tab.
36. _____ Notice that the left side of the display is a list entitled Set Colors For. This is the list of ladder window objects on which you can modify the text and background colors. To change the page title background and text colors, select Page Title. In the Text Color area, click on a new color.
37. _____ In the Background Color area, select a new color.
38. _____ Notice that the colors you selected are displayed directly below the Select Colors For list. If you are satisfied with the new colors, click on OK to return to the ladder window and view your modified ladder window.
39. _____ Return to the View Properties window and the Colors tab.
40. _____ Change the rung comments background colors and text colors.
41. _____ Click on Apply to accept the changes without leaving the View Properties window.
42. _____ Click on the Fonts tab. This tab gives you the opportunity to change the text style and size.
43. _____ Click on the Address Display tab. This tab is where the address display format is set up. The cross-reference display can be turned either on or off here. The cross-reference display will display cross-reference data on your ladder window for either inputs or outputs. A cross-reference simply lists other locations on your ladder program where the selected instruction address is also used.
44. _____ Check both cross-reference display options.
45. _____ Click OK to return to the ladder window. Notice the cross-reference information displayed. Information like 5:0, 10:1, and 11:2 indicates that the selected address is also used in ladder file 5 rung 0, ladder file 10 rung 1, and ladder file 11 rung 2.
46. _____ Click on the Miscellaneous tab. Experiment with each of the options as they were demonstrated by your instructor.

47. _____ Click on the Quick Key Mapping tab.
48. _____ Review Quick Mapping. Quick key mapping is another programming method. The object of quick key programming is to eliminate key strokes as you enter instructions on your ladder rungs. Quick key allows assigning of instructions to your computer's alpha keyboard keys; this is called mapping. After the keys are assigned instructions (mapped) quick key mode must be enabled. After enabling, the programmer simply presses the keyboard key associated with the instruction to place the information on the ladder rung. For example, the C key could represent one of the available counter instructions.
49. _____ Click OK to return to the ladder window.
50. _____ Experiment with expanding and shrinking different file folders by clicking on the + or − sign.
51. _____ Familiarize yourself with the contents of the Help folder. We will investigate the help available in a future lesson.
52. _____ Expand the Controller folder.
53. _____ Double-click on Controller Properties.
54. _____ Observe the information contained under the General tab.
55. _____ The Passwords tab is where a password is set up.
56. _____ Locate the Controller Communications tab. We will work with the Controller Communications tab when we work with setting up our drivers and going on-line with our processor.
57. _____ Click Cancel to return to the Project Window.
58. _____ Expand the Program Files folder. The Program Files folder stores ladder programs. LAD 2 is the main control or ladder program. The SLC 500 family of PLCs can only have one main ladder program file. All other ladder files are subprograms or subroutines. Notice other ladder files, such as LAD 3 and LAD 4. These are subroutines.
59. _____ Double-click on LAD 3. Ladder file 3 opens and displays its ladder logic.
60. _____ Double-click on LAD 4. Ladder file 4 should open.
61. _____ Double-click on LAD 2 to reopen the main ladder file or program.
62. _____ Expand the Data Files folder. The Data Files folder holds files containing all data used in conjunction with the ladder program and its subprograms. The cross-reference information we displayed in the View Properties window is stored in the Data Files folder in report form. The Cross Reference report can be printed. Also contained in the Data Files folder are files containing other types of data. Input data, output data, timer and counter data, along with numerical and bit data, are all stored in the data files. Notice the different folders identifying this data.
63. _____ Right-click on the output data files to observe the right-click options menu. Your display should show the right-click options similar to Figure 2-9. Familiarize yourself with the available options.
64. _____ Click off the right-click menu to return to the project window.
65. _____ Double-click on the N7 Integer data file. Your display should be similar to Figure 2-10. The Integer file stores whole numbers.
66. _____ Click on the Properties button to view the Integer file's properties screen. We will work with different parts of this and other data file screens when we investigate addressing and data files in a later lesson.
67. _____ Click Cancel to return to the Integer file.
68. _____ Close the Integer file.

Other Files in the Data Files Folder

Force Files are used to force I/O points on and off. Forcing of I/O is typically a maintenance or installation task.

The Custom Data Monitor feature provides a method to customize a display window so an individual can monitor data from different data files contained in the Data Files folder.

Trends are used to create an oscilloscope-type screen displaying operating data from your machine or process.

The Database is where the text displayed on your ladder program is stored. Page titles, rung comments, instruction descriptions, and comments are stored in the database.

3

The RSLogix 500 Help

OBJECTIVES

Upon completion of this laboratory exercise, you should be able to:

- navigate the RSLogix 500 software to find and open up various Help topics
- access topics in the Quick Start for project development help
- access Instruction Set help
- understand how to get help with programming software questions
- open Help topics for questions in the Windows 95 or NT operating environment

INTRODUCTION

There is no printed manual for the RSLogix software. All the information you need will be found on the Help screens. Help is found on the right end of the Windows Menu Bar and in the Project Tree. While working through this exercise, you will open various Help screens to familiarize yourself about where to go when you need help with the programming software, the Windows operating environment, or the SLC 500 family instruction set. Remember, if you are using RSLogix 500 newer than version 4.5, your screens may look different.

EQUIPMENT REQUIRED

You will need a personal computer running RSLogix 500 software.

INSTRUCTOR'S DEMONSTRATION

Follow along as your instructor demonstrates navigating through the Help screens.

THE LAB

1. _____ Have your computer running, and open the RSLogix 500 software.
2. _____ Open the "begin" project.
3. _____ Click on the Help selection on the Windows Menu Bar. You should see the drop-down menu as illustrated in Figure 3-1.

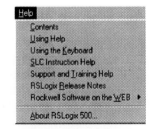

Figure 3-1 RSLogix 500 Help selections.

Help is divided into the following sections:

Contents

The first Help section is divided into four areas: Quick Start for Project Development, Programming Reference and Overview, Understanding the Operating Environment, and Detailed Look at the Complete Instruction Set. The Quick Start is a really handy section that can be used to point you in the right direction regarding programming questions or procedures. See Figure 3-2.

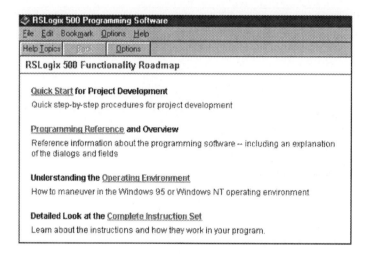

Figure 3-2   The Contents option is divided into four sections.

Using Help

If you're having trouble using the software's Help screens, this section can provide some help. Arranging and opening windows, setting colors, copying, finding definitions, setting bookmarks, and annotating a Help topic are a few subjects covered in this section. When selecting the Using Help option, a window will open with three tabs across the top. These tabs are labeled Contents, Index, and Find. Both the Index and Find tabs allow you to enter a key word to help you search for the help you need.

Using the Keyboard

Information on Windows hot keys and shortcuts will be found in this section. Using the keyboard section is similar to using Help. When selecting the Using Keyboard help option, a window will open with three tabs across the top. These tabs are Contents, Index, and Find. Both the Index and Find tabs allow you to enter a key word to aid in the search for the help you need.

SLC Instruction Set Help

The SLC 500 Instruction Set help provides detailed information on each instruction used in RSLogix 500 programming. When you click on this selection, a window opens that provides two methods for getting help on a specific instruction. The top portion of the screen contains a table with a number of boxes; each box contains an instruction written as a three-letter abbreviation called its mnemonic. If you know the mnemonic, simply click on it to go to that specific instructions Help screen. Figure 3-3 illustrates the top half of the SLC Instruction Set Help window. Notice the arrangement of boxes and their associated mnemonics.

If you are unfamiliar with the mnemonics, proceed to the bottom of the screen to the Categories of Instructions (see Figure 3-4). Here instructions are identified as Math Instructions, or

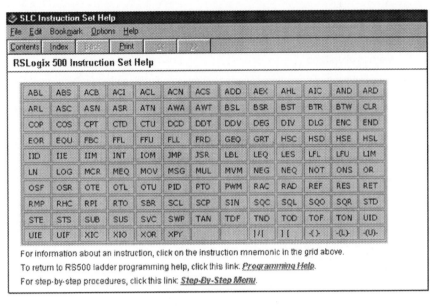

Figure 3-3   Instruction Help mnemonic help.

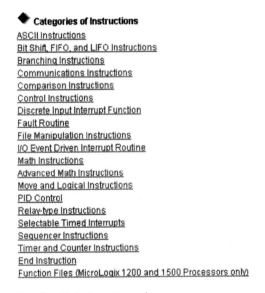

Figure 3-4   SLC 500 Instruction Set Help by categories.

Timer and Counter Instructions, for example. Simply click on the specific group of instructions relating to the topic for which you need assistance. Notice that the instructions are grouped in categories.

Clicking on the Math Instructions category will display a window that has a list of all instructions in that group. Select the instruction from the "If You Want to . . ." description of the instruction. The instructions help screen will be displayed. Figure 3-5 illustrates a portion of the Math Instructions help screen.

Notice in Figure 3-5 that if you want to add two values, you select the ADD instruction. For this instruction, ADD is the mnemonic. Clicking ADD will display the addition instruction help.

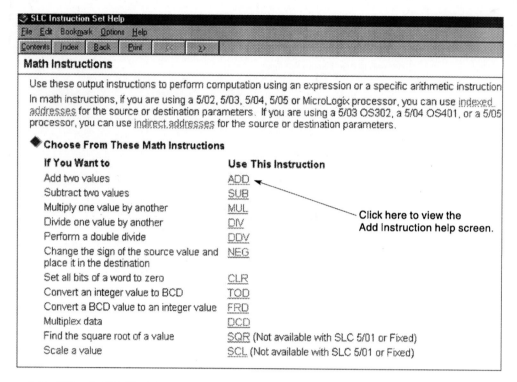

Figure 3-5   RSLogix 500 Math Instructions help screen.

Both methods take you to the same point. They merely provide different ways to get there, depending on your experience with the software.

Support and Training Help

For information on how to contact technical support, obtain product updates, renew software support, and learn about training options available through Rockwell Software, use Support and Training Help.

RSLogix Release Notes

Release notes for the software provide information about each time the software is upgraded and made available to customers. This upgrade is called a release. The version of RSLogix 500 software used when developing this lab manual was Version 4.5. The notes accompanying the release of this product to customers provide the user with information on:

- updating RSLogix 500
- current enhancements
- current repairs
- known anomalies

Typically, the release notes also contain release information for all past versions of the software.

Rockwell Software on the Web

This section provides access to Rockwell Software support on the Internet. The three sections are: Internet Support, Product Updates, and the Support Library. Internet access is needed to use this feature.

About RSLogix 500

This section contains information regarding the software revision, the serial number, and to whom the software is registered. Phone numbers for Rockwell Software information are also available at this location.

4. _____ Let's investigate the Contents section of the help screens. From the Help drop-down menu, select Contents. The RSLogix 500 Functionality Roadmap window opens as illustrated in Figure 3-6.

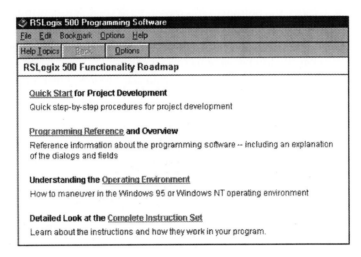

Figure 3-6   The main selection of Help Topics when clicking on the Contents option.

5. _____ Click on Understanding the Operating Environment to open it.

6. _____ You will see a number of topics such as Using Scroll Bars, Using Drag and Drop, and Splitting the Viewing Window. Refer to Figure 3-7. Open up a number of these and review the information contained there.

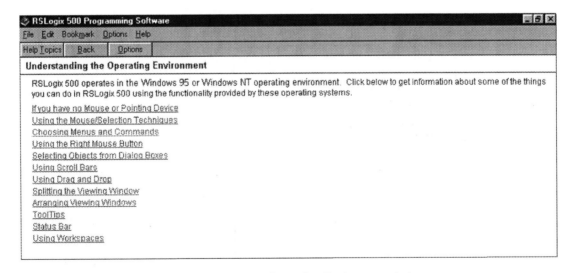

Figure 3-7   Options under Understanding the Operating Environment help.

7. _____ When completed, click on the Back button near the top left of your screen to return to the RSLogix 500 Functionality Roadmap window.

8. _____ Open the Programming Reference and Overview topic.

9. _____ Here you will see over 40 programming reference topics. Take a minute to scroll through the options. Familiarize yourself with the topics listed here so you can return when you have questions as you work through later programming exercises.

10. _____ Under the section Managing the Display Environment, open the Toolbars and Icons selection. It outlines how to view and what is contained on the standard toolbar.

11. _____ Notice the bottom of the screen has a section called Related Topics. Click on each of the four selections, one at a time, to view information on each toolbar.

12. _____ While reviewing information on one of the toolbars, click on the Options button near the top left of your screen.

13. _____ Two interesting selections from the Options button are Print Topic and Annotate. The Print Topic button will allow you to print the topic you are currently viewing. This provides you the opportunity to create your own hard copy manual of topics of interest to you.

14. _____ If you have the Instruction Toolbar topic open, click on Annotate. The Annotate feature provides you with the ability to add your own notes to a help screen.

15. _____ The Annotate dialog box should open.

16. _____ Type in a note you might want to leave for yourself regarding this screen.

17. _____ When completed, click Save.

18. _____ A small paper clip should appear in the upper left-hand corner of your screen next to the first line of text for this topic. If you click on the paper clip, your notes should appear.

19. _____ After you review your notes, click Delete to remove them.

20. _____ When you have completed reviewing each of the toolbars, click on Back twice to return to the RSLogix 500 Functionality Roadmap window.

21. _____ Click on Complete Instruction Set.

22. _____ You should see Figure 3-3 at the top of your screen, while Figure 3-4 should be at the bottom. Clicking on any of the boxes in the table will display help for the associated instruction.

23. _____ Click on the ADD instruction. You should see the help screen for the addition (ADD) instruction. Review this screen to familiarize yourself with the type of information available.

24. _____ Click Back to return to the RSLogix 500 Instruction Set Help screen.

25. _____ Under Categories of Instructions, click on Math Instructions. This should display a screen similar to Figure 3-5.

26. _____ In the section "If You Want to . . ." click the ADD instruction. This should display the same addition (ADD) instruction help displayed by selecting the mnemonic from the table.

27. _____ Click Back to return to the Math Instructions screen.

28. _____ Click Back to return to the RSLogix 500 Instruction Set Help screen. There is a very important note in the center of the screen. The note reads: "Depending on the processor you are programming, not all instructions are available. Help for each instruction specifies which processor(s) the instruction may be used with." PLC instructions are processor operating system dependent. Refer to the textbook, chapter 10, which discusses PLC processors.

29. _____ Click Back to return to the RSLogix 500 Functionality Roadmap screen.

30. _____ Click on Quick Start for Project Development.

31. _____ Notice the topics listed under "What do you want to do?".

32. _____ Under Create a Project, click on Open a project. A window titled "Open an RSLogix 500 Project" should display on the right. Refer to Figure 3-8. Notice the underlined words within the text. Under number 2, the words "document" and "folder" are both underlined. If readers are not familiar with the underlined word, they can click on it to get help understanding the term.

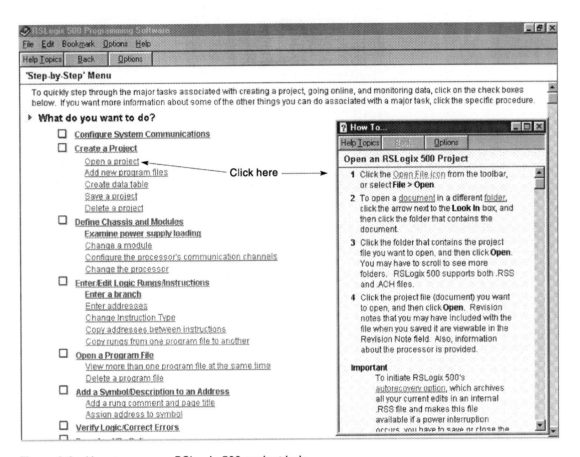

Figure 3-8   How to open an RSLogix 500 project help.

33. _____ Scroll to the bottom of the Open an RSLogix 500 Project window. Notice the following:

- The section entitled Important
- Click on this book icon for more details
- What do you want to do next?

Each of these provide additional information. The "What do you want to do next?" feature tries to anticipate your next move, and offers help.

34. _____ Near the top of the Open an RSLogix 500 Project window, click on the Options button. Note the options available, especially Annotate and Print Topic.

35. _____ Click on a blank area not on the drop-down list to return to the step-by-step menu.

36. _____ Click on Help Topics near the top left portion of your screen. A window similar to Figure 3-9 should display. Figure 3-9 displays a window with three tabs. Clicking on the Contents tab provides you with the options displayed in Figure 3-9. Click on Display Environment to select it. Click on Open to display the Help associated with the Display Environment. Clicking Print Topic will print the topic.

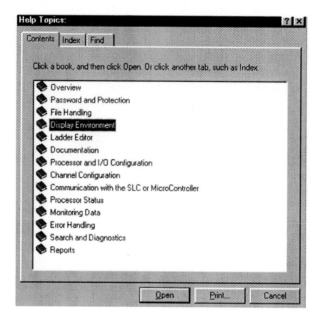

Figure 3-9 Help Topics window Contents tab with Display Environment help selected.

37. _____ Click the Index tab.
38. _____ Assume we want help on data files. Refer to Figure 3-10.
  A. Near the top of the window, type in Data Files (number 1 in Figure 3-9).
  B. The Data Files topic is displayed (see number 2).
  C. Click on the Display button (see number 3).

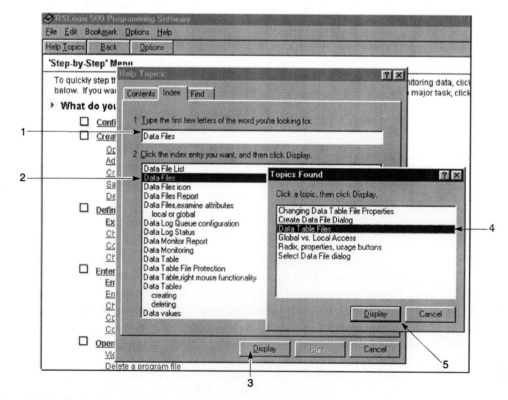

Figure 3-10 Topics Found window located under the Index tab.

D. Select the topic you desire. Here Data Table Files is selected (see number 4).

E. Click on the Display button (see number 5).

F. The Data Files help screen will be displayed.

39. _____ Click on File and Exit in the upper left corner of your screen to exit Help. This should return you to your main project window.

This completes the lab exercise introducing the RSLogix 500 Help. We introduced you to the help screens and how to navigate through them in this lesson. You need to spend time exploring the help screens on your own, as there is a wealth of information available to assist you with questions as you work with the software.

LAB EXERCISE

4

Working With Binary Numbers

OBJECTIVES

Upon completion of this laboratory exercise, you should be able to:

- use and apply binary numbers as they are used in a PLC
- convert binary numbers to decimal
- convert decimal numbers to binary

INTRODUCTION

PLCs use various number systems for data representation. Depending on your particular PLC, the current application will dictate which number systems will be used. Figure 4-1 lists the number systems used in the SLC 500 PLC family.

Number System	Radix	Characters	Common PLC Usage
Binary	2	1, 0	Data representation inside PLC
Octal	8	1, 2, 3, 4, 5, 6, 7	Data Highway Plus station addresses and Remote I/O addressing
Decimal	10	1, 2, 3, 4, 5, 6, 7, 8, 9	SLC 500 family addressing
Hexadecimal	16	1, 2, 3, 4, 5, 6, 7, 8, 9, A, B, C, D, E, F	Used to mask data in selected data manipulation instructions

Figure 4-1   Common numbering systems used in PLCs.

The most common number systems used in SLC 500 processors are binary and decimal. Unless you are using a 5/04 PLC connecting to a Data Highway Plus network where the station or node addresses are octal, or remote I/O, also addressed using octal addresses, you may never use octal numbers in an SLC 500 system. Hexadecimal numbers are used in masking. We will investigate masking later in this manual.

This exercise will help you to understand binary numbers and how they are used in the PLC 5 or SLC 500 family of PLCs. Answer the following questions:

1. As humans, we understand and use the _decimal_, base _10_, number system.

2. Microprocessor-controlled devices such as PLCs and variable-frequency drives, as well as digital computers, use the _binary_ number system.

3. What is a bit? _A binary number 1 or 0_

4. Where does the term "bit" come from? _Binary digit_

30

5. What is a computer word as used in an SLC 500 or MicroLogix PLC? _16 bits_

6. Define LSB. _Least significant bit_

7. Define MSB. _Most significant bit_

8. What position in a binary number holds the LSB position? _Right most bit_

9. What position in a binary number holds the MSB position? _Left most bit_

10. What is the sign bit? _0 positive 1 negative to represent a number being pos. or neg in the left most position._

11. In what position in a 16-bit word does the sign bit reside? _Left_

12. What is a 16-bit signed integer? _____

13. What is the data range of a 16-bit signed integer? _-32,768 to +32767_

14. What is a 16-bit unsigned integer? _A 16 bit integer with the leftmost digit being pos or neg 0 1_

15. What is the data range of a 16-bit unsigned integer? _32767_

16. The SLC 500 family of PLCs uses what integer data type? _Binary_

17. Most current PLCs use 16-bit data words. Newer technology PLCs use _32_ -bit data words.

The table shown in Figure 4-2 illustrates the format and weighting of binary numbers.

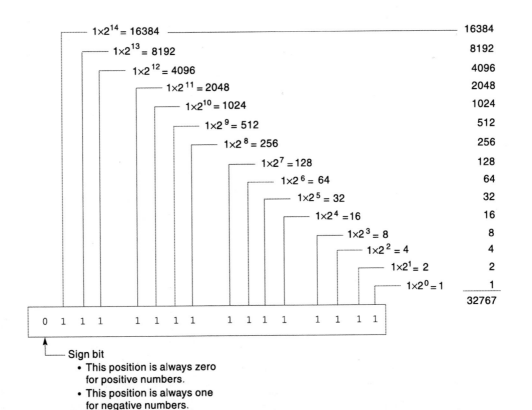

Figure 4-2 Binary number weighting.

Using Figure 4-2, convert the binary value 00000000 11010101 to a decimal value. The easiest way to accomplish this is to add up the place values as illustrated in Figure 4-3.

Place Value	Binary Value
128	1
64	1
32	0
16	1
8	0
4	1
2	0
1	1
Total	213

Figure 4-3   Converting a binary value to decimal.

Exercise

1. From memory convert the following binary numbers to decimal:

   1010 = __10__   1110 = __14__

   1101 = __13__   0111 = __7__

   0110 = __6__   1111 = __15__

   10001 = __17__   0011 = __3__

   1011 = __11__   11010 = __26__

   11100110 = __486__   1100110000101 = __6536__

2. Convert the following decimal numbers to binary:

   37 __100101__   295 _____

   13 __1101__   1234 _____

   79 __1001111__   14863 _____

3. The number 100_{10} is equal to what value in decimal? __100__

4. The value 100_2 is equal to what value in decimal? __4__

Binary to Decimal Conversion Application Scenario

This exercise will illustrate the need for a programmer, electrical worker, or maintenance worker to understand binary numbers and their conversion to decimal values.

Application Overview

The scenario for this application is a remote waste-water pumping station. Remote pumping stations are not manned and can be located quite a distance from the control station. In many cases, the pumping station may be controlled by an SLC 500 PLC and a variable-frequency drive. The SLC controls the drive, which controls the speed of the pump motor. The SLC 500 processor communicates back to the control station through a radio frequency modem. The radio frequency modem is connected to the serial port of a 5/03 or 5/04 SLC 500 modular processor. The PLC program controls the operation of the pumping station. A 4- to 20-milliamp (mA) analog signal is input into the PLC to tell it how fast the drive should run the motor due to current water conditions. Since this is a critical application, if for any reason the 4- to 20-mA speed signal is lost as an input to the PLC, the PLC will be directed to command the variable-frequency drive to pump at full speed. The PLC will send a signal to the control station regarding the situation.

INTRODUCTION TO THE LAB

As PLCs mature and obtain more features, users discover the days of setting DIP switches on I/O and specialty modules are fading fast. Many newer modules have no DIP switches. The DIP switch configuration has been replaced by software configuration. The module will be configured either as part of the RSLogix 500 project's I/O configuration or as separate configuration words and sent to the module by way of the output status file. For this example, we will use a Spectrum Controls 1747sc-INI4i four-channel, isolated analog input module. Spectrum Controls is a Rockwell Automation/Allen-Bradley partner. Being a partner, Spectrum Controls builds PLC hardware for PLCs like the SLC 500, PLC 5, or ControlLogix. The "sc" in the part number string identifies this as a Spectrum Controls module.

After the module is installed, the module must be configured by setting up a binary configuration word for each channel. Since the 1747sc-INI4i is a 4-channel analog input module, configuration data for each channel must be sent to the module. This configuration word must be sent to the output status table word and transferred to the module during an initialization subroutine, or in conjunction with the first pass bit (S:1/15) and either a move or copy instruction in the main ladder file. Figure 4-4 illustrates a ladder rung with the first pass bit and a move instruction.

Figure 4-4   Move instruction sending the value 135 to configure an analog channel at address O:4.0.

Figure 4-5 shows a copy instruction used to configure an analog module at the same address. The move instruction can send one configuration word per instruction. The copy instruction can send multiple configuration words. In this example the copy instruction has a length of 4. Four words will be sent to the output address. This copy instruction will configure a four-channel analog card with one instruction. The copy instruction can send up to 128 words for an SLC 500 or PLC 5 family processor. Four move instructions would be needed to accomplish the same thing.

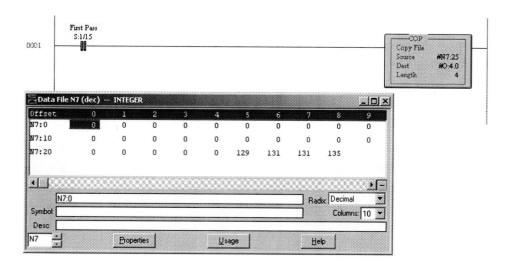

Figure 4-5   Copy instruction used to configure a four-channel analog module at address O:4.0.

Notice the source parameter for the copy instruction is N7:25. This is the address where the channel configuration information will be found. Since the instruction's length parameter is four, four consecutive words will be transferred when the instruction is executed. The integer file, N7, is illustrated and words N7:25, N7:26, N7:27, and N7:28 will be transferred to configure the four channels. We will look closer at the ladder logic used to configure this type of module a little later in this manual.

The binary configuration word will define the characteristics of each of the module's channels. As an example, bits within the configuration word will be set up for enabling the channel, selecting the input voltage or current range, and determining how input data is scaled and how the module is to behave if the analog input signal is lost.

Since this is a four-channel analog input module, one configuration word for each channel will need to be determined and sent to the module.

THE LAB

Configure a 1747sc-INI4i SLC 500 analog input module's input channel configuration word. Module configuration specifications and an explanation are as follows:

A. Enable the input channel

Each channel must be enabled or told to turn on.

B. 4- to 20-mA input range

Input signal is 4 to 20 milliamps.

C. Data format to be proportional counts

The data formatted into the PLC input status table will be in the format of −32,678 to +32,767.

D. Pump to run on maximum output in the event of an open analog input circuit

Drive to run full speed if input signal lost.

E. 60-Hz input filter

Filtering for noise.

F. Auto-calibration disabled

Module is capable of auto-calibrating itself. Disable for this application.

G. Unused bits

These bits are not defined. If not set to zero, a configuration error will occur.

Figure 4-6 represents the 16-bit data word options for the data word that needs to be programmed in either a Move or Copy instruction in the RSLogix 500 ladder program and sent to the analog module for channel configuration. Use the application information provided to construct the configuration word for the review questions.

Questions

1. What is the 16-bit channel configuration word for this application? _____
2. After determining the binary bit pattern, the value needs to be converted to decimal. The decimal value will be programmed on a ladder rung in a Move or Copy instruction. The instruction will send a copy of the information to the analog input module to configure it by way of the output status table. What decimal value would you program into your PLC Move instruction to configure this analog channel? _____

INPUT CHANNEL CONFIGURATION WORD															
Use these bit settings in the channel configuration word															
To select . . .	15	14	13	12	11	10	9	8	7	6	5	4	3	2	1	0
Configuration Bit 0, Input Channel Enable															
Input channel disable																0
Input channel enable																1
Configuration Bits 1–3, Input Range															
±10 Vdc input range (-INI4vi only)													0	0	0	
1–5 Vdc input range (-INI4vi only)													0	0	1	
0–5 Vdc input range (-INI4vi only)													0	1	0	
0–10 Vdc input range (-INI4vi only)													0	1	1	
0–20 mA input range													1	0	0	
4–20 mA input range													1	0	1	
Invalid													1	1	0	
Invalid													1	1	1	
Configuration Bits 4–6, Data Format															
Engineering units										0	0	0				
Scaled for PID										0	0	1				
Proportional counts										0	1	0				
1746-NI4 compatible format										0	1	1				
User-defined scale A										1	0	0				
User-defined scale B										1	0	1				
Invalid										1	1	0				
Invalid										1	1	1				
Configuration Bits 7 and 8, Open Circuit Response															
Zero on open input circuit[1]								0	0							
Max. on open input circuit[1]								0	1							
Min. on open input circuit[1]								1	0							
Invalid								1	1							
Configuration Bits 9 and 10, Input Filter Frequency															
60-Hz input filter						0	0									
50-Hz input filter						0	1									
150-Hz input filter						1	0									
500-Hz input filter						1	1									
Configuration Bit 11, Auto-Calibration															
Auto-calibration disabled					0											
Auto-calibration enabled					1											
Configuration Bits 12–15, Undefined bits															
Unused	0	0	0	0												
Invalid	1	1	1	1												

[1] Applies only to the 1–5 Vdc and 4–20 mA input ranges.

Figure 4-6   Spectrum Controls' input channel configuration word. (Table data compiled from *SLC 500 Isolated Analog Input Modules Owner's Guide* [Bellevue, WA: Spectrum Controls, 1998], 29.)

5

Connecting Your Personal Computer and Your PLC

PREREQUISITE

Before completing this lab, read textbook Chapter 3, Programming a Programmable Controller.

OBJECTIVES

Upon completion of this laboratory exercise, you should be able to:

- configure an RSLinx driver using a personal or industrial computer with a 1784-KTX(D) card
- set up an 1747-PIC driver to communicate with a DH-485 PLC processor or isolated link coupler
- configure an RSLinx driver using a notebook computer with a PCMCIA card
- set up a serial communication driver using RSLinx between a computer and PLC
- set up an Ethernet communications driver for an SLC 5/05 Ethernet processor

INTRODUCTION

RSLogix 500 software is used to develop and edit ladder programs. A second software package, RSLinx, is needed to set up communication between the PLC processor and the personal computer.

In order to monitor PLC activity, upload a PLC project into your personal computer, or download a project from your personal computer to your PLC, a communication link must be made between the personal computer and programmable controller. We will use RSLinx to configure our communications driver.

This lesson will step you through configuring the most common RSLinx drivers.

NEW TERMS

Upload	To transfer a copy of the PLC processor project up to a personal computer's RAM memory.
Download	To transfer a copy of a PLC project (program) from a computer or handheld programmer's memory into a programmable controller's memory.
On-line	Viewing a project in the PLC processor through your personal computer via a communication link.

Off-line To view or edit a PLC project on the hard drive of the personal computer. When off-line, there is no communication between the PLC processor and personal computer.

Driver A driver is an independent software application used to configure communications between a personal or industrial computer and a hardware device such as a PLC, variable frequency drive, or operator interface terminal. To program, edit, and monitor an SLC 500 or MicroLogix ladder file, RSLogix 500 software is used. Communication drivers are set up using a separate piece of software called RSLinx.

RSLinx RSLinx Lite is the communication driver software bundled with RSLogix. An RSLinx driver is configured to allow communication between the personal computer and the PLC processor.

SLC 500 NETWORKS

Networks are becoming increasingly popular in the manufacturing environment as a method of passing information between devices on the plant floor. Information can also be passed between the plant floor and upper levels of the company and even onto the Internet.

There are many reasons to connect plant floor intelligent devices on a network. Some examples are as follows:

- network plant floor devices together to share process information
- network plant floor devices to personal computers in the engineering office to allow PLC monitoring and program editing without transporting a computer to the plant floor
- incorporate operator interface devices (HMI) into the manufacturing process to enable operators to input information into the process as well as view information on process status from a central touch-screen terminal
- provide production information from the plant floor to the production supervisor's personal computer as an Excel spreadsheet

There are many networks available today, depending on the complexity of your system, type of data to be shared, distance data are to be transported, and the amount and speed of data transfer.

The SLC 500 family of PLCs has two basic direct connect networks, Data Highway-485 (DH-485) and Data Highway Plus (DH+). Even though there are many more complex and feature-rich networks available such as Ethernet, Control Net, and Device Net, we will introduce the two networks most likely to be used in a classroom environment as they pertain to configuring and using RSLinx.

Data Highway-485 Network

Remote I/O provided the ability to distribute remote chassis around the factory floor by connecting them back to the processor chassis with communication modules. A network provides the opportunity to make a single chassis PLC part of a larger factory-wide communication network. Each PLC on the network will have its own CPU. With a CPU in each chassis, better control of each portion of the process can be obtained with a local control CPU rather than one CPU communicating with multiple remote chassis over a slow serial communication connection.

The Data Highway-485 network is the basic SLC 500 family network. DH-485 provides the ability to connect MicroLogix PLCs; SLC 500 fixed PLCs; and the 5/01, 5/02, 5/03 processors, in addition to operator interface devices such as Panel Views, on one network. Figure 5-1 illustrates a small DH-485 network. Notice that each device on the network has a unique identifier called a node address. Isolated link couplers are necessary to provide isolation between the network cable, called the backbone, and the processors. The isolated link couplers are also repeaters or amplifiers. The object of the network is to allow all devices on the network to communicate with each other.

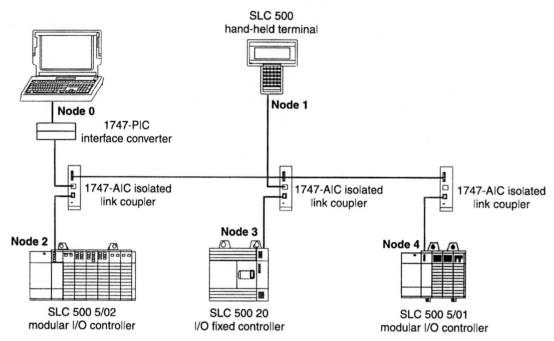

Figure 5-1 An example of an Allen-Bradley Data Highway-485 network. (Used with permission of Rockwell Automation, Inc.)

Now information can be shared between different pieces of hardware. An operator using the personal computer at node 0, or the handheld programming terminal at node 1 can program, monitor, or edit the project in any of the PLCs on the network. Figure 5-1 might simulate a manufacturing line where three PLCs control the process. Rather than have each PLC control only its respective portion of the manufacturing process, the network provides the ability for one PLC to send information, or talk, to another. If there were a jam-up on the conveyor at node 4's PLC, the node 4 PLC could send a message to nodes 2 and 3 to stop production so products would stop coming down the line. Information could also be sent to an operator viewing the personal computer at node 0. Listed below are major specifications for a DH-485 network.

32 decimal nodes, 0 through 31

Up to 4,000 cable feet

Isolated link couplers required as isolators and repeaters

Data Highway Plus Network

The most popular SLC 500 modular processor at this time is the SLC 5/04. The 5/04 processor connects directly to the DH+ network. The DH+ network is actually the network for the PLC 5. The PLC 5 is the big brother to the SLC 500 PLC family. In today's manufacturing environment there are many thousands of PLC 5 systems in use. Plants currently using the PLC 5 and its network DH+ requested an easy connection for SLC 500 processors to the DH+ network. The 5/04 was the result of that request. The 5/04 processor has two ports or connection points, called channels. Channel 1 is the DH+ connection. Channel 1 is comprised of two different connection points that are internally connected. The connection used is dictated by the interface cable. Refer to Figure 10-20 in your textbook for a picture of a 5/04 processor and its connections. Connecting to either point provides access to the DH+ network. An overview of DH+ network specifications is listed on the next page. Make special note that node addresses are in octal numbers.

Up to 10,000 cable feet at 57.6 K baud

Network speed: 57.6 K, 115.2 K, or 230.4 K baud

Node addresses are 0 to 77 octal

Total 64 nodes

How the Network Fits into Configuring RSLinx

When configuring RSLinx, users must know the type of network they are connecting to. This will narrow their interface options between the personal computer and the network. The particular interface used will dictate the particular RSLinx driver to be configured. Also, the node address must be known. When transferring information between the personal computer and a PLC processor, the node address of where the information is coming from or going to must be specified in your RSLogix 500 software before uploading, downloading, or going on-line.

HARDWARE CONNECTION

There are many ways to communicate between a PLC and a personal computer. Two things that determine the available communications options include the personal computer you are using and the specific PLC and processor to which you are connecting. Your interface could range from a direct serial connection using a null-modem serial cable, a 1747-PIC, 1784-PCMK card, or a 1747-KTX(D) card. Your connection could be a direct serial connection, through a Data Highway-485 network, or the Data Highway Plus network. The table shown in Figure 5-2 lists the most common communication interfaces between SLC 500 family PLCs and modular processors.

PLC	Serial	DH-485	DH+	Ethernet
MicroLogix 1000	X	X		
MicroLogix 1200	X	X		
MicroLogix 1500	X	X		
Fixed SLC 500		X		
SLC 5/01		X		
SLC 5/02		X		
SLC 5/03	X	X		
SLC 5/04	X	X	X	
SLC 5/05	X	X		X

Figure 5-2   Common communication interface between SLC 500 family PLCs and a network or personal computer.

A number of interface cards are commonly installable in your desktop personal computer's expansion slots. Possible interface cards include a 1784-KT, a 1784-KTX, and a 1784-KTXD. Refer to Figure 3-13 in your textbook for a picture of a 1784-KTX interface card and its connections. Refer to your textbook for additional information on interfacing options between a personal computer and a PLC.

RSLinx is the software package used as the driver between your personal computer and the PLC processor. When setting up a printer so you can print a word processor document created on a personal computer, you have probably set up a printer driver in your personal computer. The printer driver is a piece of software that allows the word processor document from your personal computer to travel through the interface cable to the printer and produce a card copy print. RSLinx works much like the printer driver for RSLogix software. RSLinx allows RSLogix to communicate through the interface cable to the PLC processor.

There are five different packages, or versions, of RSLinx. The version of RSLinx included or bundled with RSLogix software is called RSLinx Lite. RSLinx Lite, commonly called Linx Lite, provides the basic drivers required to communicate between your personal computer and PLC processor. In most cases, drivers included with Linx Lite satisfy most common connections. Linx Lite is not commercially available; it is bundled with Rockwell Software products that require direct access to RSLinx drivers.

RSLinx Packages

RSLinx Lite

RSLinx Lite is bundled with Rockwell software products that use RSLinx to communicate to plant floor devices. This package provides no external interfaces for DDE (Dynamic Data Exchange), OPC (OLE for process control), or custom "C" applications.

RSLinx OEM

The OEM package is intended for driver development with third-party human-machine interface (HMI) devices or custom applications. RSLinx OEM provides all drivers for Allen-Bradley devices and networks.

RSLinx C SDK

RSLinx Software Development Kit (SDK), provides all necessary files and documentation to develop custom C/C++ applications.

RSLinx Professional

The professional version is the full-featured version. It contains all the functionality from RSLinx OEM plus additional functionality for DDE and OPC connectivity.

RSLinx Gateway

RSLinx Gateway contains all the functionality of RSLinx Professional, plus remote client connectivity by bridging industrial networks over TCP/IP.

RSLinx SCREENS

Before we begin working with RSLinx, let's look at the screens you will see displayed on your computer and identify the major parts of each. When opening RSLinx with a DH+ driver loaded, the screen shown in Figure 5-3 will appear. Main screen pieces are identified as follows:

A. Header identifies version of RSLinx
B. Display RSWho
C. RSLinx to continuously (Autobrowse) browse the network
D. Configure drivers
E. Driver diagnostics
F. Tree Control pane
G. List Control pane
H. Data Highway Plus driver network icon
I. Driver displayed in Tree Control pane
J. Display List Control pane in detail view
K. Help
L. Display List Control pane in large icon view
M. Refresh network browse if Autobrowse is not checked

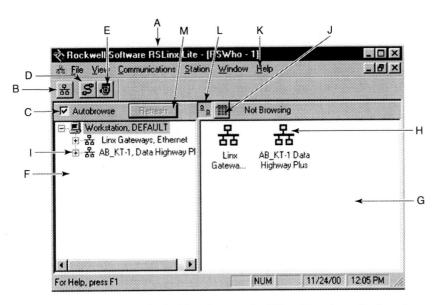

Figure 5-3   RSLinx screen with a default driver icon and a DH+ driver icon displayed.

The RSWho Screen

RSWho is RSLinx's network browser. The browser allows you to view all active network nodes. The RSWho screen also has two panes. The left pane is the Tree Control. The Tree Control pane shows the configured networks and devices. The List Control pane is on the right. Either double-click on your driver (H) or click on the + sign on the driver in the Tree Control pane (I) to display the devices on your driver's network. The List Control pane will display network information in one of two formats. Click on (L) from Figure 5-3 to see the large icon view, or (J) for the detail view. Double-click on (H) to display the network nodes. Currently, Figure 5-4 is displaying RSWho in the large icon view. RSWho in Figure 5-4 shows a personal computer named RSLinx at node 00, an SLC 500 modular PLC at node 01, a second SLC 500 PLC at node 04 (B) with the name SLC_1 (C), and a Panel View operator terminal named PV600C at node 02.

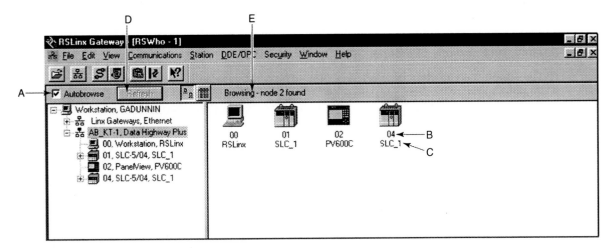

Figure 5-4   RSWho showing network nodes.

Figure 5-4, the RSWho screen, is in Autobrowse and Refresh is grayed out (D). (A) is the Autobrowse feature button. In order to keep the RSWho screen up-to-date, RSLinx must continually browse all network processors. If Autobrowse is checked, RSLinx will continuously browse the network for the driver selected in the RSWho screen (refer to E). Continuously browsing the network generates a lot of network traffic, which could contribute to bogging down the network and network performance. If you need to keep RSWho open, uncheck the Autobrowse button and click on Refresh (D) when RSWho needs to be updated.

All devices in Figure 5-4 are on the DH+ network. A programmer has access to each of these devices from his or her personal computer at node 0. Keep in mind, Figure 5-4 illustrates a small DH+ network. A DH+ network can have up to 64 devices. Figure 5-5 illustrates a similar network with an additional SLC 500 PLC at node 10. If for some reason a device or node on the network that was once active stops communicating, RSLinx will identify it with an "X" through the devices icon. This is illustrated in Figure 5-5.

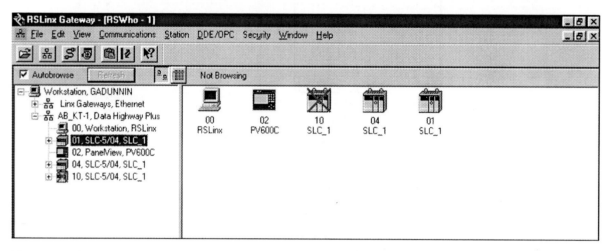

Figure 5-5   Network node 10 has stopped communicating with the network.

RSLINX DRIVER CONFIGURATION

Probably the most commonly used interface to a PLC is through a serial connection, a 1784-KTX card installed in your desktop personal computer, or a notebook computer and a PCMCIA-style card like Rockwell Automation's 1784-PCMK card. The RSLinx setup for each is explained in the following information.

Serial Connection Between a Personal Computer and a PLC

The simplest connection between a personal computer and the PLC is a serial connection. Even though simple, connecting serially can be especially frustrating if the incorrect serial cable is used. Many newer PLC processors, such as the 5/03, 5/04, and 5/05 come with RS-232 serial communication ports. When communicating between a personal computer or industrial computer and a PLC, we need the capability to download information from the computer terminal to the PLC processor. We also need to upload programs and program related-data from the PLC and the computer terminal. When data is transferred in two directions, this is called bidirectional or full-duplex communication.

It might seem that any PLC processor with an RS-232 serial port could communicate with any other RS-232 port. This is not necessarily true. When configuring a communication link between two devices, such as a computer and a PLC processor, there are two important aspects of the communication link. First, there is the communication standard.

The RS-232 Communication Standard

The RS-232 communication standard defines only the physical cable connections and use for each of the nine wires inside the standard communication cable, and their associated connector pins. Remember, when referring to a communication cable, the connector pin numbers are used for wire identification. The standard does not define how many pins and wires must be used.

Minimum configuration for two-way communication only requires three wires in the 9-pin D-shell connector. In a typical RS-232 connection, the personal computer uses pin 2 for data output while peripheral equipment, such as a modem, uses pin 2 for data input. When sending data back from the modem to the computer, pin 3 is data output from the modem; yet pin 3 is data input on the personal computer. Pin 7 is used as the ground. Figure 5-6 illustrates the minimum connections between a personal computer and peripheral equipment, like a modem. Notice that the wires go directly between the two devices pin for pin. This is referred to as a straight-through connection.

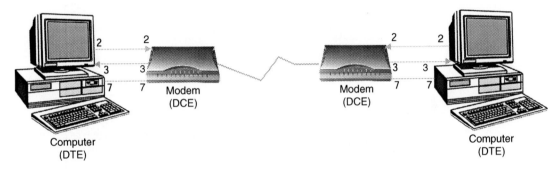

Figure 5-6   Straight-through cabling between computer and peripheral equipment.

For ease of connection, the RS-232 standard specifies that computer devices have male connectors, while peripheral equipment has female connectors.

When communicating directly between a personal or industrial computer and a PLC processor, also a computer, there is no intermediate peripheral equipment. If the same straight-through cable was used to connect the personal computer to the PLC processor (refer to Figure 5-6), we would be connecting pin 2 of one computer to pin 2 of the other computer as illustrated in Figure 5-7.

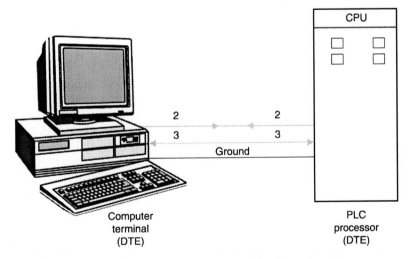

Figure 5-7   Connecting two computer devices with a straight-through cable.

A computer's pin 2 is outgoing data. Figure 5-7 illustrates both computer devices sending output data from pin 2 and attempting to use each other's pin 2 as an input for data. Both devices are looking for input data on pin 3. In this cabling configuration, pin 3 on each computer device is connected to pin 3 on the other. This type of connection will not allow communication between the two devices. Cabling needs to be modified so that the output of one computer (pin 2) is connected to the input of the opposite computer (pin 3). The output from each computer (pin 2) must cross and connect to pin 3, the input of the opposite computer. Communication sent by one computer can then be received by the other computer. Data sent back by the receiving computer can be received by the originator of the transmission. Figure 5-8 illustrates the necessary minimum connection to communicate between two computers. The common name for the communication cable illustrated in Figure 5-8, where wires 2 and 3 are crossed, is a null-modem cable.

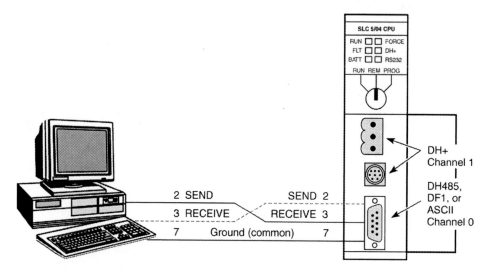

Figure 5-8 Serial wiring connections between a personal computer and channel 0 of a 5/04 processor using a null-modem cable.

The null-modem cable will have female 9-pin or 25-pin D-shell connectors on each end. It is called a null-modem cable because it replaces two modems. Each modem will be a peripheral that will enable two computer devices to communicate with each other. Make certain a null-modem cable is used when configuring a serial connection through RSLinx.

Communication Protocols

Even if two different PLC processors both support RS-232 communication, there is one other consideration when connecting two devices together: the protocol. The protocol is a set of rules that govern the way that data is formatted and timed as it is transmitted between the sending and receiving devices. Each manufacturer designs a protocol that defines data format, timing, sequence, and error checking. As a result, one processor from one PLC manufacturer will probably not be able to talk to a processor from another manufacturer, even if they both support RS-232 communication standards, as their protocols will differ. Rockwell Automation/Allen-Bradley's RS-232 communication protocol is known as RS-232 DF1. DF1 identifies the protocol.

Configuring RSLinx for Serial Interface

RSLinx is a separate software package that runs on your personal computer. RSLinx must be opened and drivers configured before communication between a personal computer and PLC using RSLogix 500 software.

NOTE: We will be assuming a default installation of RSLinx. We will be using RSLinx Lite version 2.42. If you have a different version, some of your screens may appear different even though the configuration steps will be similar. Check with your instructor for instructions if your RSLinx is installed other than default.

Personal Computer to PLC Communication Cabling

The following SLC 500 family PLC processors can be communicated with by using an RSLinx serial driver:

MicroLogix 1000 using a 1761-CBL-PM02 cable

MicroLogix 1200 using a 1761-CBL-PM02 cable

MicroLogix 1500 using a 1761-CBL-PM02 cable

SLC 5/03 modular processor using a null-modem or 1747-CP3 cable

SLC 5/04 modular processor using a null-modem or 1747-CP3 cable

SLC 5/05 modular processor using a null-modem or 1747-CP3 cable

THE LAB

Configuring a Serial Driver

1. _____ Connect the proper serial cable between your personal computer and PLC processor. The connection will be similar to Figure 5-8.
2. _____ Note which communications port you connected the serial cable to on your personal computer. Record the communications port information below.
3. _____ I used COM _____ on my personal computer.
4. _____ Power up your SLC 500 PLC.

RSLinx Configuration

1. _____ Click on your Start button in Windows.
2. _____ Click on Programs.
3. _____ Click on Rockwell Software.
4. _____ Click on RSLinx.
5. _____ Click on RSLinx. This should open the software.
6. _____ After the software opens you should see the screen shown in Figure 5-9 if no drivers are loaded.

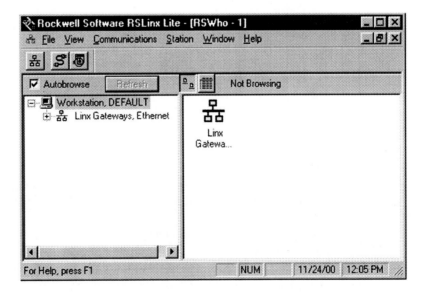

Figure 5-9   Opening RSLinx screen.

7. _____ Clicking Communications on the Windows menu bar pulls down the menu of communications commands. See Figure 5-10.

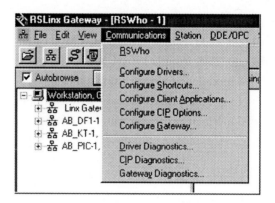

Figure 5-10 RSLinx drop-down communications menu.

8. _____ Click on Configure Drivers.
9. _____ The Configure Drivers window should open as shown in Figure 5-11.

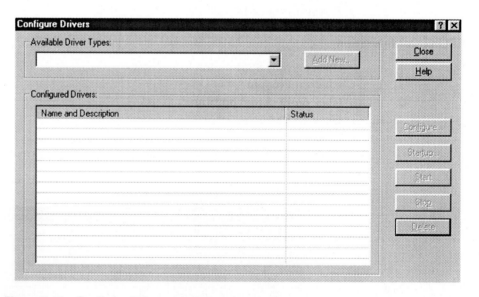

Figure 5-11 Configure Drivers window.

10. _____ The upper left of the Configure Drivers window contains the available driver types. Click on the down arrow.
11. _____ The left section of the window contains the available drivers. We will be communicating serially using an RS-232 cable. Scroll down until you see RS-232 DF1 devices. See Figure 5-12.
12. _____ Click on this selection to select it.
13. _____ Click on Add New.
14. _____ The Add New RSLinx Driver window will open with the serial default RSLinx name AB_DF1-1. See Figure 5-13. A driver name of up to 15 characters can be entered. For this exercise, click on OK to accept the default name.

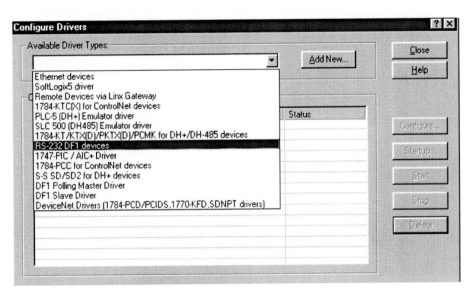

Figure 5-12   Available driver types drop-down list.

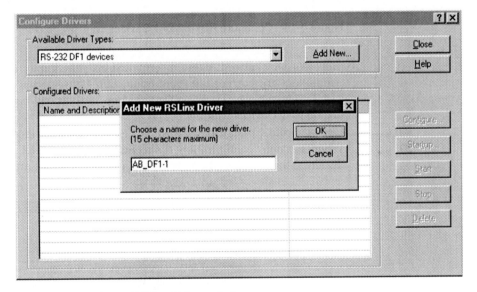

Figure 5-13   Add new RSLinx Driver window.

15. _____ The Configure RS-232 DF1 Devices window opens as illustrated in Figure 5-14.

16. _____ Click on the arrow associated with the Comm Port selection, A in Figure 5-14. The drop-down menu will list COM1, COM2, and COM3.

17. _____ Select the communications port that your serial cable is connected to on your personal computer. See step 3 under Configuring a Serial Driver.

18. _____ Device, Baud, Rate, Parity, Stop Bits, Error Checking, and Protocol selections need not be modified as we will have the software do an Auto Configure. The auto configure sequence will tell RSLinx to test each of these selections' options against the communication configuration of the SLC 500 processor we are attempting to communicate with. RSLinx will try each communication parameter setup by sending a signal to the processor, "Is anybody there?" When the communication parameters match up with the processor's channel serial communication parameters, the

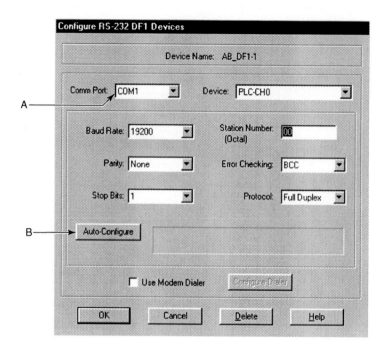

Figure 5-14   Configure RS-232 DF1 Devices window.

processor will answer, "Here I am." RSLinx now knows that the communication parameters are correct and communication between the personal computer and PLC can take place. These communication settings will be reflected in the Configure RS-232 Devices window.

19. _____ Click on Auto-Configure as shown in Figure 5-14, identified as (B).

20. _____ When the auto configuration is successful, there will be a confirmation message displayed in the box to the right of the Auto-Configure button as in Figure 5-15.

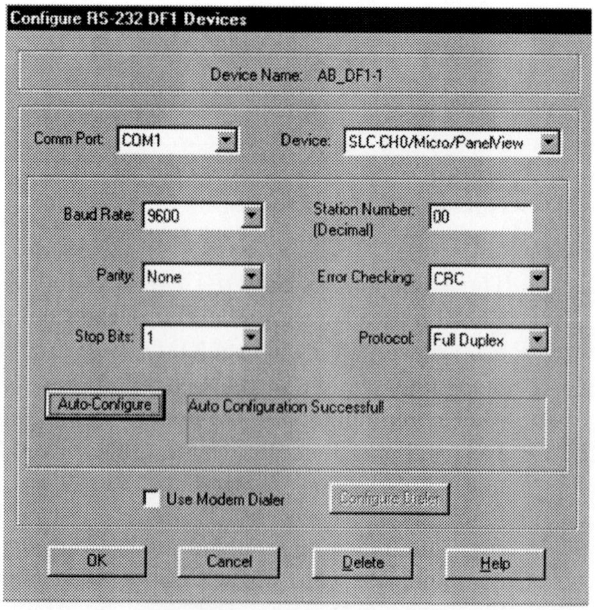

Figure 5-15   RS-232 Auto-configuration successful.

21. _____ Click on OK to return to the Configure drivers screen.

22. _____ Your newly configured driver should be listed in the configured drivers area as illustrated in Figure 5-16, which identifies what is contained in the Configured Driver Name and Description entry.

    A. Driver name. We accepted the default name, AB_DF1-1.

    B. Station 00 is the node address of your computer terminal.

    C. COM 1 is the personal computer COM port used.

    D. Driver is running.

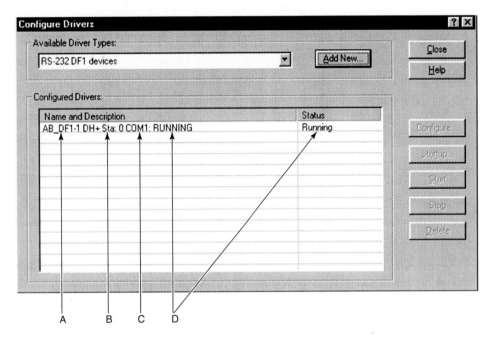

Figure 5-16   Configured AB_DF1 serial driver.

23. _____ Close the Configure Drivers Screen.

24. _____ The RSWho screen should show your drivers as illustrated in Figure 5-17.

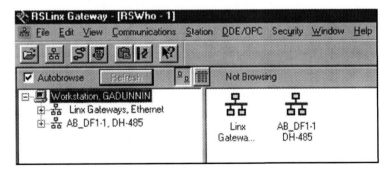

Figure 5-17   AB_DF1 serial driver icon displayed on the RSWho screen.

25. _____ Click on the AB_DF1-1 network icon. You should see a graphic of your computer terminal (A). Notice that the computer terminal is node 00. There should also be a graphic of your PLC (B). The PLC is identified as node 01. The PLC processor is always identified as node 01 when using a serial connection. A serial connection is a direct, or point-to-point connection. Since a serial connection is a direct connection

and not a network connection, RSLinx defaults the processor to node 01. Remember that your PLC is node 01 when you set up communications in the RSLogix 500 software. Figure 5-18 is an RSWho screen showing a terminal and an SLC 500 modular PLC.

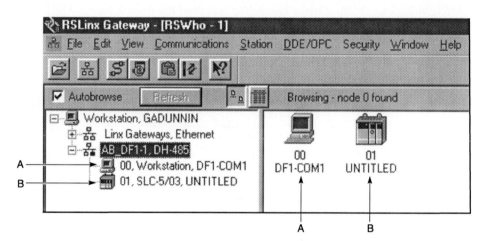

Figure 5-18   Modular SLC 500 RSWho screen showing the personal computer as node 00 and SLC 500 as node 01.

If you are using a MicroLogix 1000 PLC, Figure 5-19 illustrates an accurate RSWho screen. The personal computer is node 00 with the name DF1-COM1, while the MicroLogix 1000 is node 01 and named TEST.

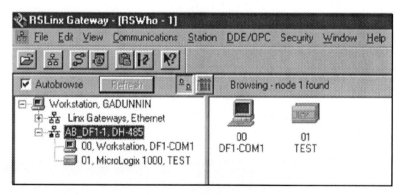

Figure 5-19   RSWho screen showing the personal computer as node 00 and MicroLogix 1000 as node 01.

A MicroLogix 1200 or 1500 will have a similar configuration with a similar icon identifying the specific product. Notice in both cases that the processor node address is node 01, while the personal computer is node 00. Remember, a node address for the computer or the PLC processor was not specified while configuring the driver. An important point to understand is that a serial connection is a direct connection, also called a point-to-point connection. When connecting serially, there is a direct connection between the personal computer and the PLC. A direct connection is not a network connection. When connecting serially between a PLC and a personal computer, there is no network connection, nor do you have the option to go on the network or see other devices on the network. This is true even if you serially connect to a PLC processor such as a 5/03, 5/04, or 5/05 that is also connected to a network.

26. _____ Minimize RSLinx on the task bar, do not click on the "X" as this will shut off RSLinx.

27. _____ This completes your configuration of a serial driver in RSLinx.

Serial Configuration Problems

If your driver has problems configuring, you will see the message in the window to the right of the Auto-Configure button as illustrated in Figure 5-20.

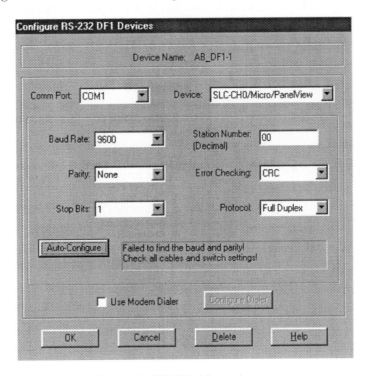

Figure 5-20   RSLinx failed to configure the RS-232 driver.

To Solve Your Configuration Problem

1. Verify that you have a null-modem serial cable, and it is connected to COM port of your personal computer.
2. Make sure the COM port selected in the Configure RS-232 DF1 devices window matches the actual COM port being used.
3. Make sure your PLC is powered.
4. From the Configure Drivers screen, click on the problem driver to highlight it.
5. Try auto-configuring the driver again.
6. If this is unsuccessful:
   - Place your PLC processor in program mode
   - Try Auto-Configure again
7. If this is unsuccessful:
   - Return to the Configure Drivers window
   - Select the problem driver
   - Delete the driver
   - Close RSLinx
   - Restart RSLinx and try again
8. If this is still unsuccessful:
   - Delete the driver
   - Shut down everything and restart your personal computer

Configuring RSLinx for a 1784-KTX Interface Card

The KTX interface card is used as an expansion card in a desktop personal computer or industrial computer expansion slot to provide interface to a PLC processor. Figure 5-21 identifies the features on the KTX card as viewed from the back of a personal computer.

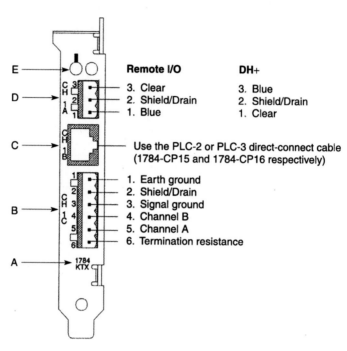

Figure 5-21 Allen-Bradley 1784-KTX Data Highway Plus interface card cable connections. (Used with permission of Rockwell Automation, Inc.)

A. Interface card identification
B. Channel 1C. DH-485 network connection. This is the network connection point for the DH-485 network connection for the 1784-KTX card shown in textbook Figure 3-14.
C. Channel 1B. Connection point for the older PLC-2 and PLC-3.
D. Channel 1A. DH+ connection to network. This connection can also be used as Remote I/O (RIO) if you are using this card as a remote I/O scanner for a SoftLogix 5 system.

Refer to textbook Figure 3-22. This is a Rockwell Automation RAC 6181 industrial computer. The 1784-KTX interface card would be placed in the shared PCI/ISA expansion slot.

The KTX interface card is also available as a 1784-KTXD. The difference between a KTX and a KTXD card is that the KTXD card had two DH+ network connection points. The KTXD card replaces the PLC-2 and PLC-3 connection point with a second RH+/RIO channel. Using this connection, a second DH+ network could be interfaced to using this card. A second RSLinx driver would need to be configured to communicate with the second DH+ network.

Configuring RSLinx

As a first step to configuring the KTX driver, start by opening RSLinx and proceed to the Configure Drivers window.

1. _____ From the Available Driver types drop-down menu box select "1784-KT/KTX(D)/PKTX(D)/PCMK/for DH+ or DH-485 devices." Refer to Figure 5-12.

NOTE: If using Windows NT, the PCMK drivers may be on a separate line in the available driver types list.

This driver group will be used to configure drivers for the following interface cards:

1784 -KT

The 1784-KT card is an older personal computer expansion-type interface to a PLC card.

1784-KTX(D)

The 1784-KTX is a half-size personal computer expansion card that must be inserted into a 16-bit ISA or EISA expansion slot. The card comes in two versions, the KTX and the KTXD. The KTX card has only one DH+ connection, or channel; the KTXD card has two DH+ channels. The KTXD card can communicate with two separate DH+ networks. Two RSLinx drivers will have to be configured, one for each network connection.

1784-PKTX(D)

This card is similar to the KTX card for general programming, configuration, and monitoring using a desktop personal computer or an industrial computer with a standard PCI bus.

1784-PKTX: one channel—DH+, RIO, or DH-485

1784-PKTXD: two channels—DH+ or RIO and DH-485

1784-PKTS: one remote I/O channel

1784-PCMK

The PCMK card is a PCMCIA type II card used to interface between a personal computer with PCMCIA slots, typically a notebook computer, and DH+, DH-485, or RIO.

2. _____ Click on the Add New button.
3. _____ The Add New RSLinx Driver window will open. The default name, AB_KT-1, will be displayed. It is suggested that you accept the default driver name.
4. _____ Click on OK.
5. _____ The window entitled Configure 1784-KTX/KTXD should open as illustrated in Figure 5-22.

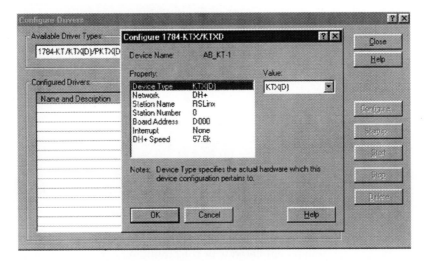

Figure 5-22   1784-KTX/KTXD configuration screen.

A. Click on the Value: drop-down arrow (A). The available selections will be displayed (see Figure 5-23).
B. Select KTX(D) from the list.
C. This will select the Device Type.
D. Click on Network and select either a DH+ or DH-485 network.
E. Click on Station Name to give your personal computer terminal a name other than the default RSLinx. This name will be displayed on your RSWho screen.
F. Click on Station Number to assign your personal computer terminal a station or node address other than the default of 0.
G. Click on Board Address to enter the KTX interface card's board address. The board address is the card's physical memory address for the expansion memory area in your personal computer's system memory. This memory is used to exchange data between the personal computer and the PLC and to operate the interface card. A minimum of 4 K bytes of memory is necessary to operate the card. The factory setting for the KTX card is D700. The KTXD card comes preset from the factory at D700 for channel 1 and D600 for channel 2. There are two rotary switches for memory address allocation for each channel of the KTX(D) card.
H. You can leave the Interrupt setting at none. Better performance can be obtained by selecting an unused interrupt such as 3, 4, 5, or 7. An interrupt is necessary if using the KTX card as an I/O scanner in a SoftLogix 5 system.
I. DH+ Speed or DH-485 Speed is the baud rate of the network you are configuring RSLinx to communicate with. The baud rate is set up in your PLC processor's Channel Configuration section of the RSLogix 500 software. All devices on the network must communicate at the same speed.
J. When you have completed setting up the properties, click on OK to configure the driver. When the driver is configured, the Configure Drivers screen should look similar to Figure 5-23.
K. If you need help along the way or are seeking additional information, click on the Help button.

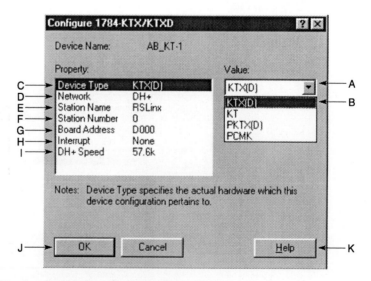

Figure 5-23   Configuring the 1784-KTX/KTXD card.

6. _____ When completed with the Configure Drivers screen, click on OK to exit. The Configure Drivers screen should look similar to Figure 5-16 with the exception that the

driver name, node address, and board address will reflect data entered for this configuration.

7. _____ When completed with the Configure Drivers screen, click on Close to exit to the RSWho screen. You should see your personal computer and PLC as in Figure 5-24. Keep in mind that Figure 5-24 illustrates an AB_KT-1 driver configured with the personal computer at node 00, with the name of RSLinx. Refer to (A) in the Figure. The SLC 500 modular PLC is at node 04 with the name of SLC_1. Refer to (B). Your RSWho will reflect your particular configuration.

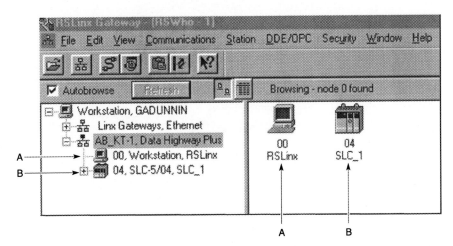

Figure 5-24   RSWho screen showing an example of an AB_KT-1 driver configured.

1784-PCMK Interface Card

Allen-Bradley has a PCMCIA card referred to as a 1784-PCMK type II card that serves as an interface between programmable controllers and personal computers used as programming terminals. The 1784-PCMK cards will interface either to Data Highway Plus network, or Data Highway-485. The PCMK card is inserted into your personal computer's PCMCIA slot as illustrated in Figure 5-25.

A series A interface cable is attached to the series A PCMK card as illustrated in Figure 5-26. The series B card and interface cable currently replaces the series A.

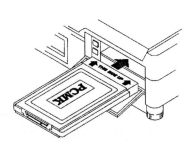

Figure 5-25   PCMK card insertion into a notebook-type personal computer's PCMCIA slot. (Used with permission of Rockwell Automation, Inc.)

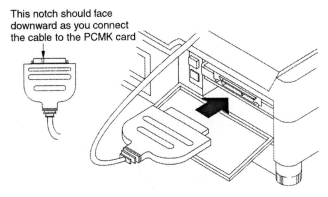

Figure 5-26   Series A interface cable attachment to a series A PCMK card. (Used with permission of Rockwell Automation, Inc.)

The cable connector is a little different; however, it connects in a similar fashion. The opposite end of the interface cable is attached to the PLC's processor as illustrated in Figure 5-27. Figure 5-27 shows the cable connections from a PCMK card to an SLC 5/01, 5/02, or 5/03 processor, or a DH-485 isolated link coupler. The cable used to connect to a 5/04 processor is a 1784-PCM5 cable and has a different connection on the processor end. This cable connects to the round connection of the DH+ channel one. Make sure you have the correct series cable for your PCMK card and the processor you are connecting to.

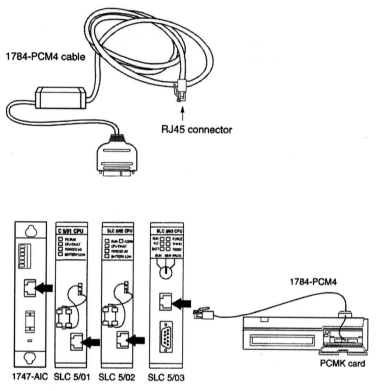

Figure 5-27   Interface cable from PCMK card for attachment to DH-485 type SLC 500 processors and isolated link coupler. (Used with permission of Rockwell Automation, Inc.)

After setting up your personal computer's RSLinx communication drivers so communication can take place between the personal computer, the PCMK card, and the PLC's CPU, you are ready to upload, download, or monitor the PLC's program on your personal computer.

Setting Up RSLinx and a PCMK Driver

Setting up a PCMK driver is very similar to setting up a KTX driver. For this exercise, we will set up the RSLinx driver to communicate to an SLC 500 5/04 processor. The 5/04 processor has two channels: channel 0, serial, and channel 1, DH+. This lab will set up the driver to communicate with the DH+ channel.

1. _____ Power up your SLC 500 PLC.
2. _____ Connect the 1784-PCM5 cable between your PCMK card and your SLC 5/04 processor.
3. _____ Open RSLinx.
4. _____ Either select Configure Drivers from the drop-down menu under Communications on the Windows Toolbar, or click on the Configure Drivers icon.

5. _____ From the Configure Drivers window, click on the down arrow in the Available Driver Types window.

6. _____ Since we are configuring a PCMK driver, we need to find and select the PCMK driver from the displayed list. This will be the same driver selected when we selected the KTX driver.

7. _____ Click on the 1784-KT/KTX(D)/PKTX(D)/PCMK for DH+/DH-485 devices. Remember, if using Windows NT, the PCMK driver may be a separate line item.

8. _____ Click Add New.

9. _____ Click OK at the Add New RSLinx Driver window. We will accept the default name AB_KT-1.

NOTE: If you have another driver configured, such as the previously set up KTX driver, the new driver's default name will be AB_KT-2.

10. _____ From the Configure 1784-PCMK window shown in Figure 5-28, click the down arrow in the Value box.

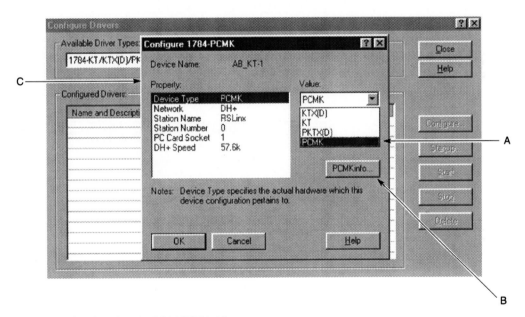

Figure 5-28   Configuring a 1784-PCMK driver.

11. _____ Click on PCMK, (A) in Figure 5-28. This selection will fill in the Device Type property in the Property window; refer to (C) in the figure.

Steps 12 through 17 pertain only if using Windows 95 or Windows 98. These steps are not necessary for Windows NT, ME, 2000, or XP.

12. _____ Click on the PCMK info box, (B).

13. _____ The PCMK info Utility window will open as illustrated in Figure 5-29.
   A. Notice that one PCMK card has been detected by your personal computer.
   B. The PCMK card is a series B card.
   C. The card is available.
   D. The card is in socket 2.

Remember this information, as some of it is required to configure the driver.

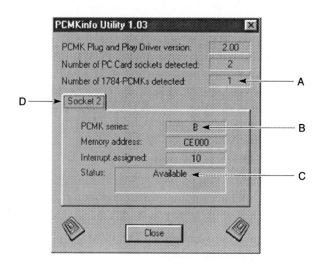

Figure 5-29   PCMKinfo Utility.

14. _____ Click on Close.
15. _____ (A) In the Configure 1784-PCMK window, click on PC Card Socket. Refer to Figure 5-30. Figures 5-30 and 5-31 will look similar when using Windows NT or later. PC card socket information, (A) in Figure 5-30, will not be displayed.

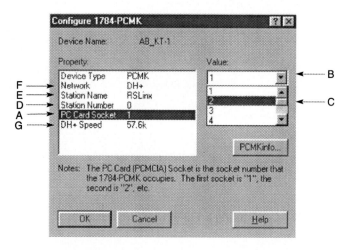

Figure 5-30   Configuring the PCMK.

16. _____ (B) Click on the drop-down arrow in the Value box.
17. _____ (C) Select the socket the PCMK card is in.
18. _____ (D) If you wish to change the personal computer station or node address, click on Station Number. Enter an unused node address in the Value box. Do not press the Enter key.
19. _____ (E) To change the personal computer name, click on Station Name. Enter the new name in the Value box. Do not press the Enter key.
20. _____ (F) If configuring a DH-485 network instead of DH+, click on Network. Use the down arrow key in the Value box to display DH-485. Click on DH-485. Do not press the Enter key.
21. _____ (G) If the DH+ network speed is other than default of 57.6 K baud, click on DH+ Speed. Select the speed out of the Value box. Refer to Figure 5-31.

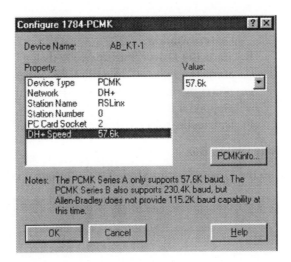

Figure 5-31   PCMK DH+ Speed configuration and notice.

NOTE: Notice the note under the property box. If you have a series A PCMK card, your only communications speed option is 57.6 K baud. With a series B PCMK card, 57.6 K baud along with 230.4 K baud are available. Remember, all devices on the network must communicate at the same speed.

22. _____ When completed configuring the Configure 1784-PCMK window, click OK.

23. _____ This should return to the Configure Drivers screen. Under Configured Drivers, Name and Description, you should see entries similar to the following, depending on your specific selections for the options in the Configure 1784-PCMK screen. See Figure 5-32 for identification of the driver information. Notice the vertical row of buttons on the right side of the Configure Drivers window.

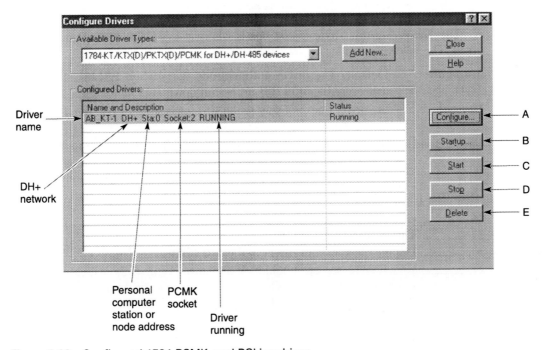

Figure 5-32   Configured 1784-PCMK card RSLinx driver.

A. The Configure button is used to configure the selected driver in the Configured Drivers window.
B. Starts the Driver Startup Mode dialog box for the selected driver.
C. Starts the selected driver in the Configured Drivers list.
D. Stops the selected driver in the Configured Drivers list. A driver cannot be stopped if it is in use.
E. Deletes the selected driver from the Configured Drivers list only if it is not running.

Driver Startup Mode Dialog Box

Clicking on the Driver Startup Mode button brings up the dialog box displayed in Figure 5-33.

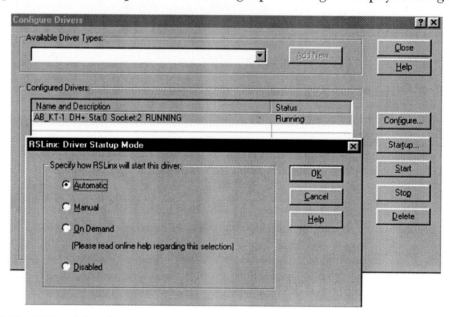

Figure 5-33   RSLinx Driver Startup Mode dialog box.

The Driver Startup Mode dialog box is where you set the startup states for the selected driver. The following options are available:

Automatic	(Default) The driver automatically starts when RSLink starts. The driver remains running until RSLinx is shut down.
Manual	The driver does not start when RSLinx is started. Use the Start/Stop buttons on the Configured Drivers screen to control the driver.
On Demand	The driver starts and stops when the application using it needs to. This selection is only available for Serial, Ethernet, and the PCMK driver when used in Windows 95 or 98. See Help screens for additional information.
Disabled	Driver does not start. Use Startup mode to change status so driver can be used.

24. _____ Close the Configure Drivers window.
25. _____ This returns you to the RSWho window.
26. _____ Double-click on your driver in the List Control screen.
27. _____ Your RSWho screen should look similar to Figure 5-24, which illustrates an RSWho screen from the RSLinx Gateway software. The RSWho screen looks the same with RSLinx Lite as well as with RSLinx Gateway. Notice the following:
  A. The personal computer terminal is set up as node 00 and named RSLinx, just as we set up in our RSLinx driver configuration.
  B. The SLC 500 PLC is node 4 and named SLC_1. The name SLC_1 is the project name that is currently in the processor.

Configuring a 1747-PIC Driver

One method to interface between a personal computer and an SLC 500 fixed, 5/01, 5/02, or 5/03 processor is to use the 1747 Personal Computer Interface Converter (1747-PIC) and configure the 1747-PIC driver in RSLinx. The PIC is not used to interface with the MicroLogix PLCs. The PIC simply connects to the serial port on your personal computer and to the DH-485 phone jack-type connection on your SLC 500 processor or isolated link coupler. Figure 5-34 illustrates the hardware interface.

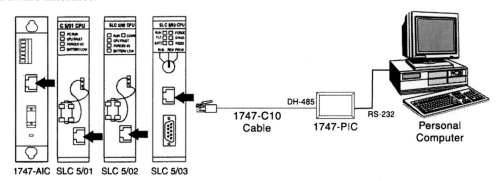

Figure 5-34 Interface cable connection from the 1747-PIC to different Allen-Bradley SLC 500 CPUs or the Data Highway 485 Isolated Link Coupler, the 1747-AIC. (Used with permission of Rockwell Automation, Inc.)

Simply connect the phone jack-type connector to your processor or isolated link coupler (AIC) as illustrated in Figure 5-34. Connect the serial connector to the serial port (COM 1 or 2) of your personal computer.

Configuring RSLinx to Communicate Via the 1747-PIC

The 1747-PIC driver cannot be configured while RSLinx is running as a service. When configuring the driver, ensure that RSLinx is running as an application.

1. _____ Open RSLinx.
2. _____ Click on the Configure Drivers icon, or choose Configure Drivers under Communications from the Windows menu bar.
3. _____ Click the down arrow on the right side of the Available Driver Types window.
4. _____ The drop-down menu as illustrated in Figure 5-35 will display.

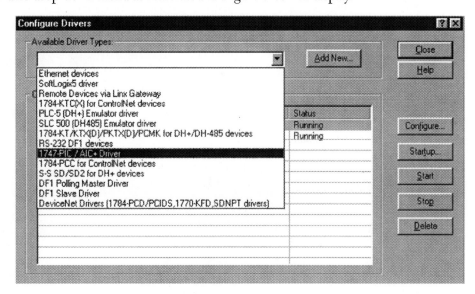

Figure 5-35 Select 1747-PIC/AIC+ Driver.

5. _____ Select 1747-PIC/AIC+ Driver, as illustrated.
6. _____ Click on Add New.
7. _____ The Add New RSLinx Driver dialog box opens as shown in Figure 5-36. You have the option to name your driver using up to 15 characters. For this lab exercise, accept the default name by clicking on OK.

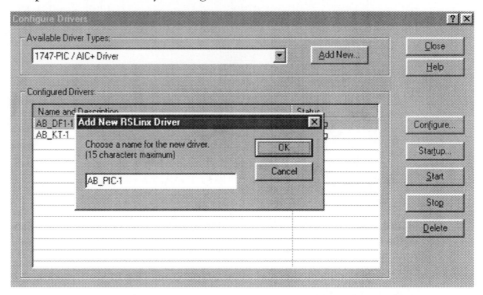

Figure 5-36 Add new driver name entry. Accept default name for this exercise.

8. _____ The Configure 1747-PIC/AIC+ Device dialog box should appear as illustrated in Figure 5-37. When configuring the driver, ensure that RSLinx is running as an application. Check that the box Reserve COM Port for Exclusive Use by this Driver is enabled. This ensures that RSLinx will work properly when you run it as a service. Refer to Figure 5-37 as you configure the driver.

   A. Station number (Decimal): Fill in the station number of the personal computer you are using. For our application, you can leave the default, 00.

   B. Max Station Number (1–31): The value placed in this area is the station address of the highest number station or node on the DH-485 network. When browsing

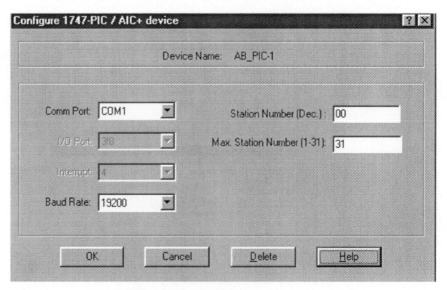

Figure 5-37 Configure 1747-PIC/AIC+ window.

the network, the PLC will not look for stations beyond this station number, thus saving network update time.

C. COM Port: Select the COM port you are using on this personal computer.

D. Baud Rate: Select the baud rate as configured in the PLC processor channel configuration.

9. _____ When completed filling in the dialog box, press OK.

10. _____ The Configure Drivers dialog box should display as shown in Figure 5-38. Notice that there are three drivers configured in Figure 5-38. The AB_DF1-1 driver is a serial driver; it is stopped. Since the serial driver and the PIC driver both share the same serial port, only one can use the serial port at a time.

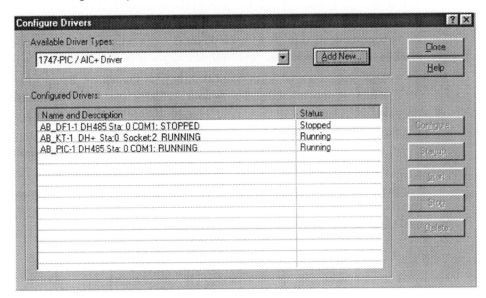

Figure 5-38 RSLinx Configure Driver dialog box with AB_PIC-1 Driver configured and running.

11. _____ Click Close.

12. _____ Figure 5-39 is the RSWho display with the PIC driver, as well as two other drivers, configured.

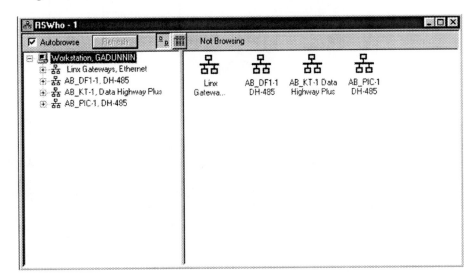

Figure 5-39 RSWho with three drivers configured.

13. _____ Double-clicking on the AB_PIC-1 icon should display a personal computer (with the name RSI-PIC at node 00 along with the PLC at node 01) with the name of the current project contained in its memory.

14. _____ Close the RSWho window.

15. _____ Minimize RSLinx to the task bar.

CONFIGURING AN RSLinx AND AN ETHERNET DRIVER

This exercise will set up an RSLinx driver to communicate with an SLC 500 5/05 processor. The 5/05 processor has two channels: channel 0, for serial, and channel 1, for Ethernet communications. Refer to Figure 5-40. This lab will take you through the steps of setting up the RSLinx driver to communicate directly with a 5/05 processor via the Ethernet channel. There are a number of steps necessary for setting up communications with an Ethernet PLC processor:

- The processor's IP address must either be known or set up to conform to the existing Ethernet network setup. The default setup for a new Rockwell Automation PLC processor is Bootp Enable.

- An Ethernet RSLinx driver must be configured.

- Any personal computer either on the network or to be installed on the network must have its IP address set up to conform to the network configuration.

For this lab, we are only going to set up the RSLinx driver portion of communications. As we proceed through Lab Exercise 10, we will complete the communications setup and perform a download of our project over Ethernet.

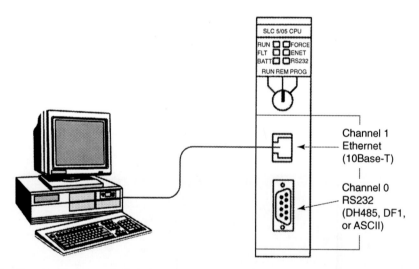

Figure 5-40 Personal computer directly connected to an SLC 500 5/05 processor.

There are two methods of communicating via Ethernet with a PLC processor: using a direct connection between a personal computer and a PLC processor, or using a network connection. Figure 5-40 illustrates a direct connection to the processor using a crossover cable. When setting up a direct connection, an Ethernet crossover cable is required.

In an industrial environment, an Ethernet network is used to connect many devices together. This connection is achieved with a series of hubs or switches that are typically installed in the enclosures (Figure 5-41). A patch cable is required between each device and the hub or switch.

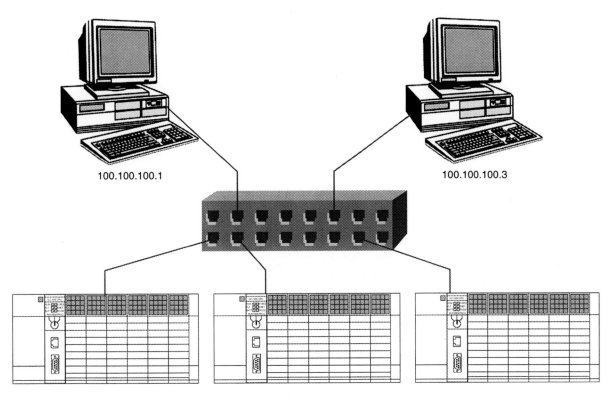

100.100.100.1

100.100.100.3

Figure 5-41   Ethernet network using a hub or switch.

The Lab

1. _____ Open RSLinx.
2. _____ Click on the Configure Drivers icon.
3. _____ Click the drop-down arrow on the right side of the Available Driver Types window.
4. _____ The drop-down menu illustrated in Figure 5-42 will display.
5. _____ Select the Ethernet devices Driver as illustrated.
6. _____ Click on Add New.

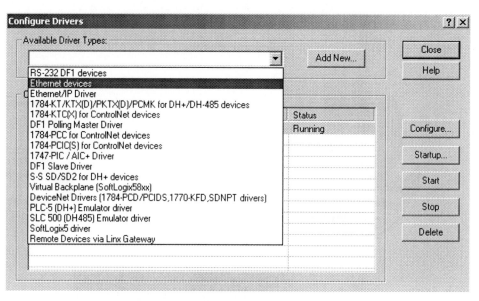

Figure 5-42   Select the Ethernet devices driver.

7. _____ The Add New RSLinx Driver dialog box opens (Figure 5-43). You have the option to name your driver using up to 15 characters. For this lab exercise, accept the default name by clicking on OK.

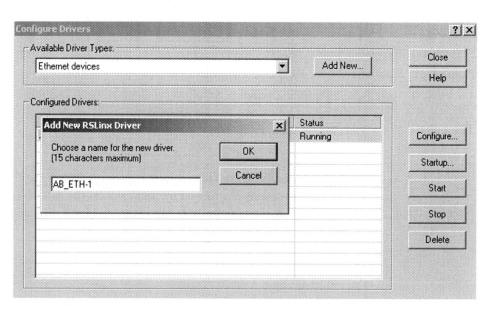

Figure 5-43   Add new driver name entry. Accept default name for this exercise.

8. _____ The Configure driver: AB_ETH-1 dialog box should appear as illustrated in Figure 5-44.

9. _____ In the box below Host Name, type the SLC 5/05 processor's IP address. For this lab, use the IP address 100.100.100.2. The dialog box should display as illustrated in Figure 5-45. Note the station number. This is the node or station you will select when you set up your RSLogix 500 software communications.

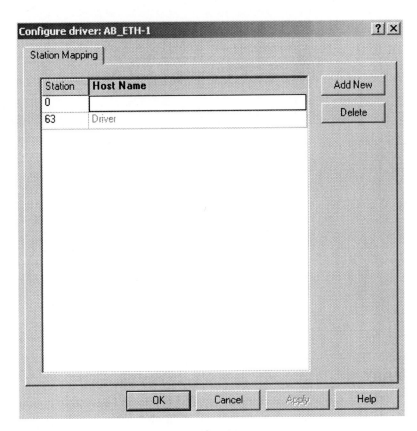

Figure 5-44   Configure Ethernet driver dialog box.

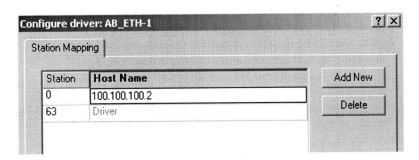

Figure 5-45   Ethernet IP address entered.

10. _____ Click OK.
11. _____ The Configure Drivers screen should display and show the driver as running (Figure 5-46). If you do not see the driver as running, have your instructor assist you in setting up your driver again.
12. _____ Click on Close.
13. _____ This completes the setup of the RSLinx Ethernet driver. At this point, because the processor channel is not set up, we will be unable to go to RSWho and see our processor. We will set up the Ethernet channel for our 5/05 processor as part of Lab Exercise 10.

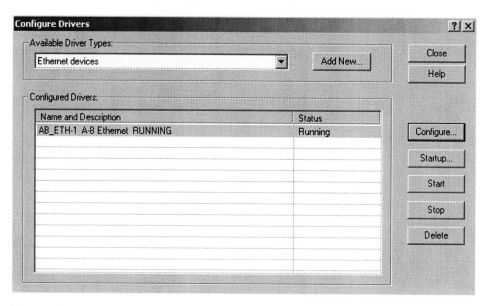

Figure 5-46   Configured Ethernet driver is running.

SUMMARY

This lesson introduced setting up RSLinx drivers to enable communication between a personal desktop or notebook type personal computer. Keep in mind that on the factory floor, industrially hardened computers called industrial computers are often used in place of the typical desktop personal computer. An industrial computer can be connected to a PLC processor much the same as can a desktop personal computer. Driver setup would be the same in both cases. Refer to your textbook for additional information on industrial computers and monitors.

RSLinx Lite is bundled with the RSLogix software you will use shortly to develop SLC 500 or MicroLogix ladder programs, called projects. After RSLinx drivers are configured, you need to go to your RSLogix software and set up communications to match the RSLinx setup. We will look at setting up RSLogix communications in a later lesson.

Once you have your RSLinx driver set up, RSLogix 500 communication parameters must be configured.

As a reminder, when you begin to configure the communications in RSLogix software, make a note of your RSLinx driver name and node address.

My RSLinx Driver name is _____.

My PLC node number is _____.

REVIEW QUESTIONS

1. What is a driver? _____

2. Define node or station address. _____

_____

3. How many nodes can there be on a DH+ network? _____

4. How are Data Highway Plus nodes addressed? _____

5. A serial connection is also called a direct connection or _____ connection.

6. A serial connection, being a direct connection, does or does not allow you to also connect to a DH1 network using an SLC 5/04 processor? Explain why or why not. _____

_____

_____

_____

7. A serial connection will always assign node address _____ to the processor.

8. When connecting serially, what type of serial cable would you select? _____

9. In what type of computer would you use a PCMK card? _____

10. In what type computer would a KTX(D) interface card be used? _____

11. What SLC 500 processors could you communicate with using the 1747-PIC? _____

_____

12. Where would you use a 1747-CP3 cable? _____

_____

13. What RSLinx driver is used to communicate to a MicroLogix 1000? _____

14. Explain the difference between a null-modem cable and a straight-through cable. _____

_____

_____

15. How many nodes are allowed on a DH-485 network? _____

16. What is the maximum cable length for a DH-485 network? _____

17. What is the maximum cable length for a DH+ network? _____

18. Is there anything that dictates maximum cable length on a DH+ network? _____

19. Which network uses an isolated link coupler? What is its function? _____

_____

_____

20. DH-485 node addresses are in what format? _____

6

Developing Ladder Logic from Functional Specifications

PREREQUISITE

Before completing this exercise, read chapter 6 in the textbook.

OBJECTIVES

Upon completion of this laboratory exercise, you should be able to:

- develop PLC ladder rungs using instructions in series
- develop PLC ladder rungs using instructions in parallel
- develop PLC ladder rungs using instructions using series and parallel input instructions
- develop PLC ladder rungs using output instructions in parallel

INTRODUCTION

Historically, the ladder diagram has been the traditional method for representing electrical sequences and operations for controlling machinery or equipment. The ladder diagram is accepted as the industry standard for providing control information from the designers to the users charged with equipment installation, modification, and maintenance. When the programmable controller was developed and introduced to industry, one of the requested features was that it also represent circuit control with the traditional ladder diagram format. With traditional ladder logic built into the PLC, maintenance and electrical personnel could easily adapt to the new technology because they were familiar with ladder logic. Even though there are new higher-level languages available today, most PLCs are still programmed with the old familiar ladder logic.

Rather than relying on electrical continuity as in a conventional circuit, the PLC uses logical continuity. The PLC implements the hardware wiring interconnections in the PLC's ladder program. In Figure 6-1, notice that inputs are wired separately into the PLC's input section. Field output devices are also wired individually to the PLC output section. The ladder program contained inside, and solved by, the CPU resides inside the PLC.

LADDER LOGIC INSTRUCTIONS

Ladder logic has conditions that need to be met before a desired action can take place. The conditions are typically programmed on the left side of the ladder rung, while the action is programmed on the right. The conditions are called input instructions, and the actions are called output instructions.

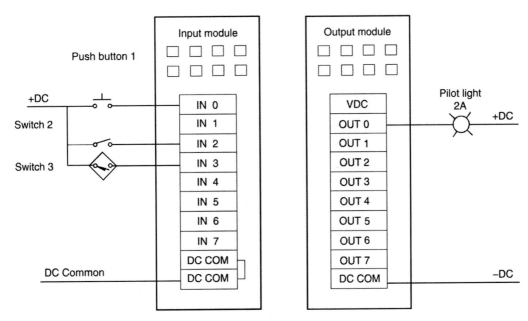

Figure 6-1   Inputs and one output wired separately into PLC I/O modules.

Instructions in Series

If *all* input conditions need to be met in order to take action, the condition is called series or AND logic. Figure 6-2 illustrates a ladder rung where push button 1 AND switch 2 AND switch 3 must all be closed before the output pilot light 2A will turn on. The wiring in Figure 6-2 is represented in Figure 6-1.

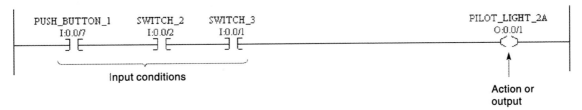

Figure 6-2   Ladder input instructions arranged in series.

Since PLC ladder logic has logical continuity rather than actual power flow, we say the instructions are either true or false. If all series input instructions are true, the rung is considered true and the output will also be true or energized.

Instructions in Parallel

Parallel ladder instruction construction states that any one of the conditions or all of the conditions must be true before the rung will become true. Figure 6-3 illustrates parallel logic, which is also known as OR logic. Push button 1, OR push button 2, OR push button 3 must be true before the output pilot light 2A will be true. Notice that if any combination of the inputs is true, or if all inputs are true, the rung will also be true.

Parallel Outputs

Outputs can also be programmed in parallel if more than one output instruction is to be true when input conditions are satisfied. Figure 6-4 shows three outputs in parallel. If input conditions are true, all three outputs will also be true.

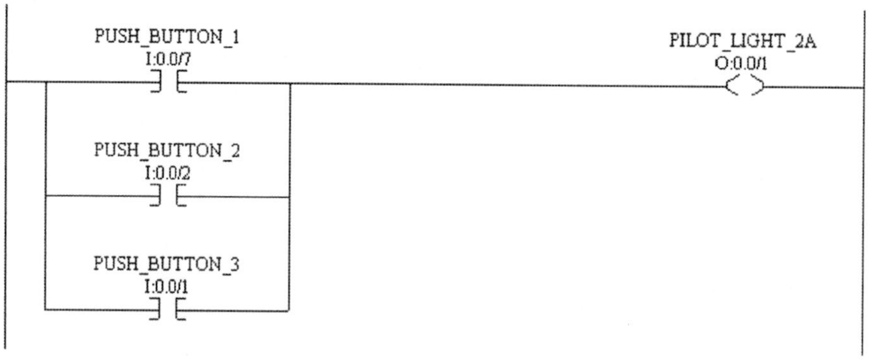

Figure 6-3   Parallel input logic.

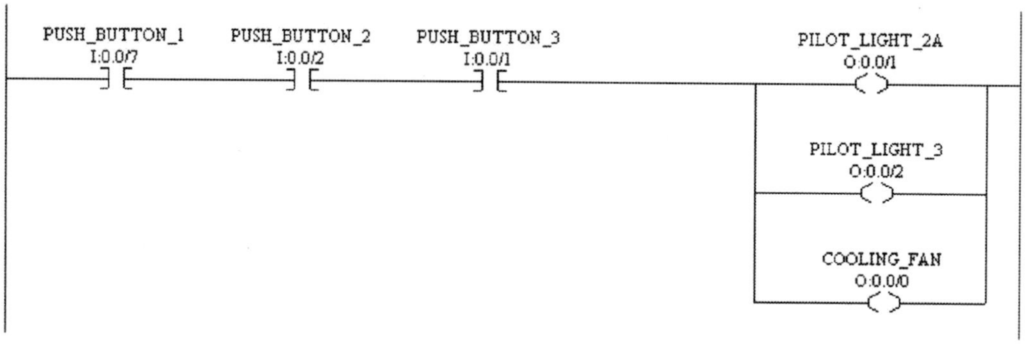

Figure 6-4   Ladder rung where three outputs will be true when the rung is true.

Combination Input and Output Logic

Instructions can be programmed in a combination of series and parallel when multiple logical input paths must be true before an output becomes true. The top rung in Figure 6-5 shows that push button 1 OR push button 2 must be true, and push button 3 must be true, before the rung will be logically true and energize the output pilot light 2A. The second rung has series logic

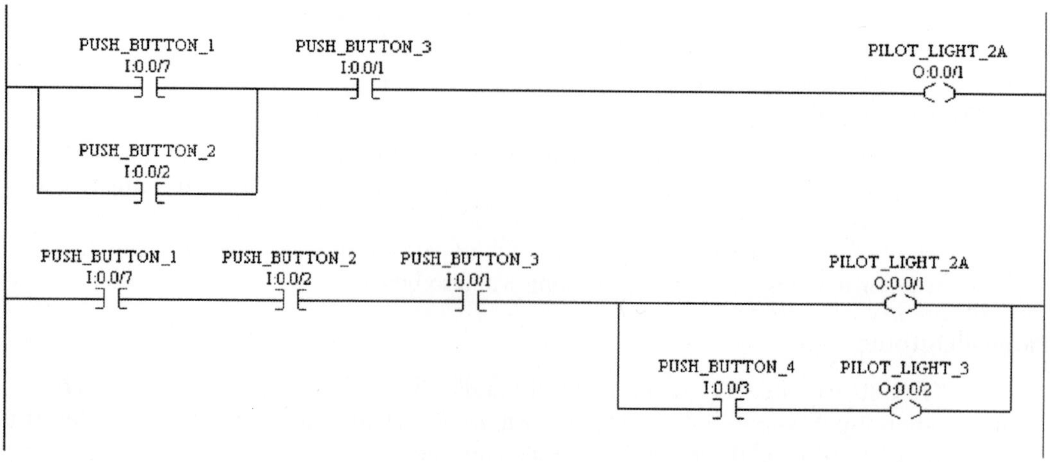

Figure 6-5   Combination input and output ladder logic.

where push buttons 1, 2, and 3 must be true to energize pilot light 2A. Before pilot light 3 will energize, push buttons 1, 2, 3, AND 4 must be true. The parallel output branch has its own input condition that must be satisfied before its output will become true or energize.

RSLogix Instruction Types

All PLCs use the common normally open and normally closed relay type instructions on PLC ladder rungs. Each manufacturer has its own names and abbreviations, or mnemonics, for these instructions. RSLogix software refers to the normally open instruction as examine on, or examine if closed (XIC). The table in Figure 6-6 describes the rules of the XIC instruction.

Examine ON or XIC Instruction	
Description: Input instruction that examines one bit for an ON condition	
If the input is	The XIC instruction is
ON or True represented by a 1 in the data table	TRUE
OFF or False represented by a 0 in the data table	FALSE

Figure 6-6 RSLogix 500 Examine-If-Closed instruction.

RSLogix software refers to the normally closed instruction as examine off, examine if open, or XIO. The table shown in Figure 6-7 describes the rules of the XIO instruction.

Examine OFF or XIO Instruction	
Description: Input instruction that examines one bit for an OFF condition	
If the input is	The XIO instruction is
ON or True represented by a 1 in the data table	FALSE
OFF or False represented by a 0 in the data table	TRUE

Figure 6-7 RSLogix 500 Examine-If-Open instruction.

The output instruction is called output energize or OTE. The OTE instruction controls one bit of information. The table in Figure 6-8 illustrates how the OTE instruction operates on your ladder rung and describes the rules of the OTE instruction.

Output Energize or OTE Instruction	
Description: Output instruction that becomes true when the rung is true	
If the rung is	The instruction will be
TRUE	True and send a 1 to the address programmed for the instruction
FALSE	False and send a 0 to the address programmed for the instruction

Figure 6-8 RSLogix 500 output Energize instruction.

Converting from Relay Ladder Diagram to PLC Ladder Logic

The following examples illustrate the conversion from a standard relay ladder diagram to PLC ladder logic. The ladder rungs in the figures are actual rungs from the RSLogix 500 software.

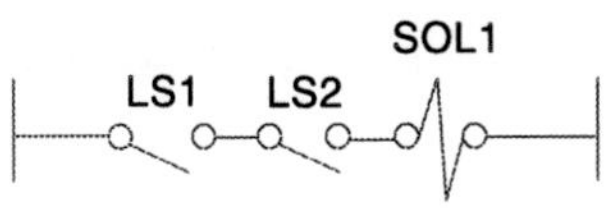

Figure 6-9 Relay Ladder Diagram for example 1.

Example 1: Two series limit switches controlling solenoid 1 are shown in Figures 6-9 and 6-10.

```
LIMIT_SWTICH_1        LIMIT_SWITCH_2                                      SOLENOID_1
   I:0.0/7               I:0.0/2                                            O:0.0/1
    ] [                   ] [                                                ( )
```

Figure 6-10   Converted RSLogix 500 ladder rung with MicroLogix 1000 I/O address.

Example 2:   Conversion of combinational input logic controlling pilot light 1 to PLC ladder format is shown in Figures 6-11 and 6-12.

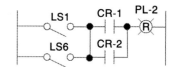

Figure 6-11   Relay Ladder Diagram for example 2.

```
LIMIT_SWTICH_5              CR_1                                       PILOT_LIGHT_1
   I:0.0/7                 I:0.0/8                                        O:0.0/1
    ] [                     ] [                                            ( )

LIMIT_SWITCH_6
   I:0.0/2
    ] [
```

Figure 6-12   Converted RSLogix 500 ladder rung with MicroLogix 1000 I/O address.

Example   3:   Conversion of parallel input logic controlling pilot light 2 to PLC ladder logic is shown in Figures 6-13 and 6-14.

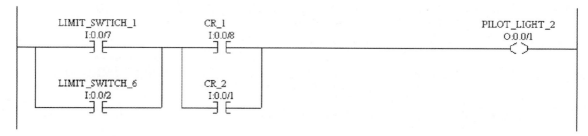

Figure 6-13   Converted parallel input logic controlling pilot light 2 to PLC ladder logic.

```
LIMIT_SWTICH_1              CR_1                                       PILOT_LIGHT_2
   I:0.0/7                 I:0.0/8                                        O:0.0/1
    ] [                     ] [                                            ( )

LIMIT_SWITCH_6             CR_2
   I:0.0/2                I:0.0/1
    ] [                     ] [
```

Figure 6-14   Converted RSLogix 500 ladder rung.

DEVELOPING LADDER LOGIC

I/O is a common term used to represent input and output. In any control situation, there are controlling signals coming into a control panel while decision signals go out of the panel to the controlled device. These incoming signals are called *inputs,* while outgoing decision signals are called *outputs.*

The first task that needs to be completed before you can develop a PLC user program is either to develop a ladder diagram that is compatible with PLC symbology, or upgrade a current conventional schematic so it can be used to develop a PLC user program.

1. Determine Real-World I/O and Allocate Addresses.

    When converting a conventional ladder diagram schematic to one in which we can develop our user program, the first task is to determine inputs and outputs that are connected to real-world hardware devices. After real-world I/O have been determined, allocate each I/O point a valid input or output address.

2. Internal Reference Allocation.

    Once the real-world I/O have been determined, internal coil instruction references need to be allocated. These internal coil instruction references are analogous to relay system relays that do not drive real-world devices but rather interact with other internal system relays. Internal coils are ladder diagram logic elements that interact with other non–real-world input or output logic internally in the CPU.

    Most PLCs allocate a portion of data memory for the storage of internal I/O statuses. If your particular PLC does not have internal coil references, you can assign unused actual output references for internal use.

3. Develop Ladder Logic in PLC Format.

4. Develop User Program.

5. Documentation of the User Program.

    A table should be developed that defines each input, output, and internal coil reference. This table should list each I/O's function. Drawings that indicate the wiring and its operation should be included for future reference. The tables should list every point available for use, even if it be for a future use. A table should be prepared for internal PLC data storage addresses. This table should be filled in during the development of the program. Documentation of your user program will be covered in the section on documentation.

DEVELOPING LADDER RUNGS FROM FUNCTIONAL SPECIFICATIONS

Before you can develop successful ladder programs you must be able to translate functional specifications into PLC ladder rungs. As an example, take the following specification:

When push button PB2 is pressed, pilot light PL2A will turn on.

Now develop a rung of PLC ladder logic from this specification. Figure 6-15 illustrates a rung of PLC logic where the push button is represented as a normally open contact (XIC) instruction and the output pilot light is represented by an output coil (OTE) instruction.

Figure 6-15   PLC ladder rung developed from functional specification.

For each of the following functional specifications, develop the correct PLC ladder rung.

1. When push button PB2 is pressed and switch SW2 is closed, pilot light PL2A will turn on.

2. When push button PB2 is pressed or limit switch LS1 is closed, pilot light PL2A will turn on.

3. If inductive proximity switches SW1, SW2, and SW3 all sense a target, motor M1 will start.

4. If any of the four doors of an automobile are open, the dome light will come on.
5. If limit switch SW1, and limit switch SW2 or limit switch SW3, are true, the full case is in position; energize glue gun to apply glue to case flaps.
6. If limit switches SW1 or SW2, and SW3, are true, energize solenoid SOL3A to move product into position.
7. If product is in position from the movement of SOL3A from question six above, energize outputs O:3/2 and O:3/7.
8. Input I:1/11 is a sensor that determines if there are box blanks in the feeder. If the boxes are not replenished by the operator and the feeder runs out of box blanks, the conveyor will be shut down and an alarm bell will sound.
9. A conveyor line is used to label and fill cans with a product. A bar-code reader is used to read the bar code on the can's label to determine that the proper label has been placed on the can.

   A photoelectric sensor is used to trigger the bar-code reader when there is a can in position. The bar-code reader will then read the can's bar code. If the bar-code reader fails to see a label with a bar code or sees a bad or damaged bar code, a no-read discrete signal will be sent to the PLC. Develop two rungs of PLC logic: one rung for the bar-code read trigger, and a second rung to alert the PLC of a no-read condition.
10. A variable frequency drive has four preset speeds it can run at depending on the conditions of three inputs from a four-position selector switch into our PLC. The drive is an Allen-Bradley 1336 Plus II Variable Frequency Drive with an interface card to accept 120-volt AC control signals. Input signal patterns into terminals 26, 27, and 28 determine at which preset speed the drive will run. The table in Figure 6-16 lists the conditions terminals 26, 27, and 28 need to be in to select a specific preset speed.

Preset Speed	DRIVE OPTION CARD INPUT SIGNALS FOR SPEED SELECTION		
	26	27	28
Preset Speed 1	False	False	False
Preset Speed 2	False	False	True
Preset Speed 3	False	True	False
Preset Speed 4	True	True	False

Figure 6-16 Variable-speed drive preset speed input truth table.

Figure 6-17 illustrates the target table for the four-position selector switch.

FOUR-POSITION SELECTOR SWITCH TARGET TABLE				
Position and Preset #1	Position and Preset #2	Position and Preset #3	Position and Preset #4	Switch Circuit Input to PLC
0	0	0	X	1
0	0	X	0	2
0	X	0	0	3

Figure 6-17 Four-position selector switch target table.

Illustrate an overview of our application. Include the selector switch, input module, ladder logic, output module, and drive option card wiring to terminals 26, 27, and 28.

LAB EXERCISE

7

Input and Output Modules

OBJECTIVES

Upon completion of this laboratory exercise, you should be able to:

- access the Rockwell Automation Web site
- obtain specifications and other technical documentation from the Internet about SLC 500 I/O modules

INTRODUCTION

Many times in the work environment the engineer, electrical workers, or maintenance personnel are in need of technical information regarding their PLC hardware or software. Today, as technology changes hourly, it is impossible to have a complete and up-to-date library of all technical publications required to do the job. However, every major PLC manufacturer has an Internet site that is available 24 hours a day, 365 days a year, just for this purpose.

This lesson will provide students the experience of obtaining technical information using the Internet to access the Rockwell Automation Web site, find three SLC 500 publications, and retrieve specific technical information.

EQUIPMENT REQUIRED

To complete this lab exercise you will need a personal computer with Internet access. The personal computer will need to have Adobe Acrobat Reader loaded. If you do not have Adobe Acrobat Reader loaded, it can be downloaded at no cost either from the Adobe Web site (www.adobe.com), or the Rockwell Automation Web site (www.ab.com).

THE LAB

For this lab we will act out the following scenario:

You are an electrical or maintenance individual working the night shift in a manufacturing facility. You are in need of specific technical information on a discrete SLC 500 input module, output module, and analog I/O module, along with wiring information on a MicroLogix 1000. After looking around the plant, you find that no technical documentation is available on the hardware in question. A colleague has mentioned that the Internet provides a wealth of information. To obtain the necessary information, you turn to the Internet.

Technical information is needed on the following I/O specifications and wiring:

- 1746-IA16 I/O module
- 1746-OA16 I/O module
- 1746-NIO4V Analog I/O Module
- MicroLogix 1000, part number 1761-L16BWB

1. _____ From your personal computer launch your Internet software.
2. _____ Go to www.ab.com.

Note: Internet sites are updated regularly. The steps below and screen references were current at the time of publication. They may be slightly different as you work through this lab.

3. _____ On the left-hand side of your screen find Publications Library. Expand this to view its contents.
4. _____ Locate Manuals On-line and click on it to open. See Figure 7-1.

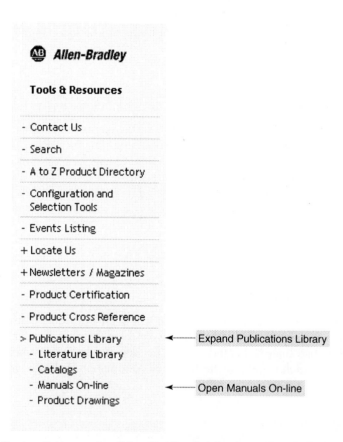

Figure 7-1 Rockwell Automation/Allen-Bradley Publications Library Navigation.

Discrete I/O Modules

5. _____ Since we are looking for information on two I/O modules, click on I/O. See Figure 7-2.
6. _____ Refer to Figure 7-3 and click on 1746 SLC I/O.
7. _____ Locate "Discrete Input and Output Modules," publication number 1746-2.35-JUL99.
8. _____ To open the publication, click PDF in the far right column.
9. _____ Wait while your computer loads the publication.

LITERATURE LIBRARY

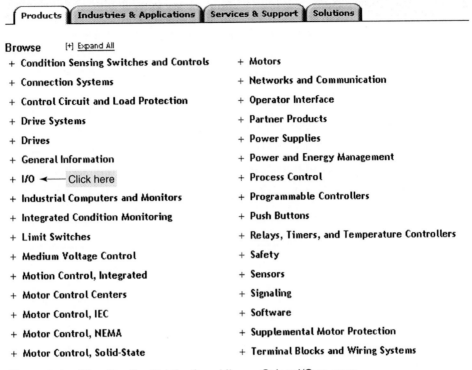

Figure 7-2 Allen-Bradley Publications Library. Select I/O to open.

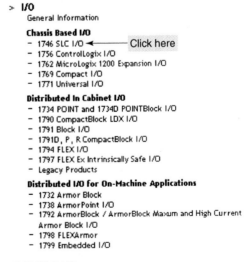

Figure 7-3 Click on 1746 SLC I/O.

10. _____ After the desired publication has been loaded, you have two options. The publication can be viewed on the Internet, or you can perform a "save as" and save the publication to your hard drive for future reference. Ask your instructor what to do here.

11. _____ Once on your hard drive, the publication, or any part of it, can be printed to provide a hard-copy reference.

1746-IA16 Module

The following questions refer to finding the technical information needed to complete the wiring of the 1746-IA16 module in the exercise scenario. The beginning of the publication has a section on terms. Review the terms to familiarize yourself with common terms associated with the I/O modules. Reviewing the publication, then locate and fill in the information needed for the 1746-IA16 module.

12. _____ Is this an input or output module? _____

13. _____ List the voltage category of this module. _____

14. _____ How many points per common? _____

15. _____ What does RTB mean under the part number? _____

16. _____ Record the maximum off-state voltage of the module. _____

17. _____ Record the maximum off-state current of the module. _____

18. _____ Record the minimum on-state voltage of the module. _____

19. _____ What is the maximum signal delay listed? _____

1746-OA16 Module

The following questions refer to finding the technical information needed to complete the wiring of the 1746-OA16 module in the exercise scenario. The beginning of the publication has a section on terms. Review the terms to familiarize yourself with common terms associated with the I/O modules. Reviewing the publication, then locate and fill in the information needed for the 1746-OA16 module.

1. _____ Is this an input or output module? _____ List the operating voltage. _____

2. _____ What is the maximum continuous current per point? _____

3. _____ What is the recommended surge suppression for a 120 Vac inductive load? _____

4. _____ Review the section on Surge Suppression. What I/O devices does the publication suggest require additional surge suppression? _____

   _____

5. _____ What is the definition of Continuous Current Per Point? _____

   _____

6. _____ Review the wiring diagrams for the module.

7. _____ If you have printer access, print a hard copy of the module wiring.

8. _____ To return to the Publication List screen, click the Back button.

Finding Technical Information for the 1746-NIO4V Analog I/O Module

1. _____ Scroll down to Analog I/O Modules.

2. _____ Locate "Analog I/O Modules for the SLC 500 Programmable Controllers," publication number 1746-TD001A-EN-P.

3. _____ Click on the underlined part number.

4. _____ Wait while your computer loads the publication.

5. _____ After the desired publication has been loaded, view the publication while on the Internet, or perform a "save as" and save the publication to your hard drive for future reference.

6. _____ Once saved to the hard drive, the document, or any part of it, can be printed to provide a hard-copy reference.

7. _____ As you locate information on the 1746-NIO4V Analog I/O Module, fill in the information needed in statements 8 through 15.

8. _____ To understand how to interpret the part number and what each digit means, locate the "Overview of Analog Modules" topic and define each piece of the module part number. _____

_____

_____

_____

9. _____ This module converts analog input signals to _____ values for storage in the SLC 500 processor's _____.

10. _____ For a 0 to 10 Vdc input signal, what is the decimal input data that will be sent to the data table? _____

11. _____ If the input signal is to be 0 to 20 milliamps, what is the decimal input data that will be sent to the data table? _____ What is the nominal resolution? _____

12. _____ If the input signal is to be 4 to 20 milliamps, what is the decimal input data that will be sent to the data table? _____

13. _____ List the module's input voltage ranges. _____

14. _____ List the module's input current ranges. _____

15. _____ What is the output range of the 1746-NIO4V module? _____

16. _____ Review the wiring diagrams for the module.

17. _____ If you have printer access, print a hard copy of the module wiring.

MicroLogix 1000 Technical Specifications and Wiring

The following questions refer to finding the technical information needed to complete the wiring of the module in the exercise scenario. Reviewing the publication, then locate and fill in the information needed for the MicroLogix 1000.

1. _____ Select the processor's category for information on the MicroLogix 1000.

2. _____ Locate the MicroLogix 1000 Users Manual, publication 1761-6.3.

3. _____ Review the glossary at the back of the book for definitions of terms.

4. _____ Wiring information will be found in chapter two.

5. _____ Review chapter two for wiring recommendations.

6. _____ Locate wiring information on the 1761-L16BWB.

7. _____ Carefully review the wiring for sinking and sourcing inputs. Notice that the same PLC can accept sourcing or sinking inputs if wired properly.

8. _____ Scroll through the chapter until you find the specific wiring for the 1761-L16BWB wiring diagram (Sinking Input Configuration).

9. _____ What is the off-state valid input range? _____

10. _____ What is the on-state input voltage range? _____

11. _____ List the output operating range of the module. _____

12. _____ Refer to chapter one, Installing Your Controller. Explain the breakdown of the part number for the 1761-L16BWB.

13. _____ If you are using a MicroLogix 1000, explain the breakdown of the part number for the unit you are using.

14. _____ If you have printer access, print a hard copy of the module wiring.

15. _____ When completed, log off the Internet.

8

Field Device Interface to Input Modules

OBJECTIVES

Upon completion of this laboratory exercise, you should be able to:

- locate and understand technical documentation on PLC input wiring
- better understand sinking and sourcing by wiring sensors to input modules
- connect two-wire and three-wire inductive proximity sensors to the PLC
- interface a two- or three-position selector switch
- interface a photoelectric sensor to the PLC
- wire a mechanical limit switch to the PLC input section

INTRODUCTION

Understanding input modules and field device interface can best be carried out by hands-on wiring of common input field devices to PLC input points. Field device interface to PLC inputs for a MicroLogix 1000 or modular SLC 500 is an important part of PLC training. This lab exercise will provide experience wiring common field input devices to the particular PLC you are using in your lab setup.

This exercise will be up to your instructor's discretion, and partially determined by the available field devices and PLC hardware. Suggested PLC interface would include sinking and sourcing proximity sensors, as well as two- and three-wire sensors, push buttons, selector switches, mechanical limit switches, and photoelectric sensors.

THE LAB

1. _____ Obtain wiring documentation for your particular PLC. Study the wiring diagrams so you understand how field devices interface to your particular PLC. These diagrams will serve as a reference as you work through the wiring exercises. Take time to understand the input wiring if you are using a MicroLogix 1000; it can be wired to accept either sinking or sourcing inputs if properly wired. If you are working with a modular SLC 500 PLC, different input modules will be required to interface sinking versus sourcing and 120-volt AC inputs. Familiarize yourself with proper wiring and field device module selection before you attempt to wire and apply power to any input module. If you have questions as you work through the exercise, consult your instructor.

If you need technical documentation and wiring information on your particular fixed PLC or modules, return to the Internet and retrieve the needed information as you did in lab exercise 7. To obtain information on the MicroLogix 1000 wiring, go http://www.ab.com, Manuals On Line, Control Processors, MicroLogix 1000, then locate the MicroLogix 1000 Users Manual, publication 1761-6.3. Wiring information will be found in chapter two, Wiring Your Controller.

2. _____ Use the Internet to find and review technical documentation for a specific output section, for a fixed PLC, or for specific output modules if you are using a modular PLC. It is not necessary to wire field devices to the modules as the module LEDs can be used to verify program execution. Check with your instructor for information on whether interfacing output field devices are a part of this lab exercise.

3. _____ After reviewing technical wiring information, select the first field input device to interface the PLC.

4. _____ Select the proper input module, or MicroLogix 1000, to interface with the field device selected.

5. _____ Remove power to the PLC.

6. _____ Perform the wiring.

7. _____ Refer to your wiring documentation and double-check your wiring.

8. _____ Have your instructor check your work before applying power.

9. _____ When all is safe, apply power to the PLC and the input device.

10. _____ Close the input device or place a target in front of the sensor.

11. _____ View the input screw terminal's associated LED. The LED should be illuminated when the input point sees a valid ON signal.

12. _____ Remove power to your input device and PLC.

Check with your instructor to determine which field devices will remain wired when you have completed this lab exercise. These input devices will be used to provide input signals for testing ladder programs that will be developed in future lab exercises.

13. _____ Select the next field device to interface.

14. _____ Repeat the procedure until each assigned input device has been wired and tested.

9

SLC 500 Family Addressing

OBJECTIVES

Upon completion of this laboratory exercise, you should be able to:

- understand the SLC 500 family PLC I/O addressing format
- identify SLC 500 modular chassis sizes
- identify modular SLC 500 slot assignments
- determine MicroLogix 1000 I/O addresses
- determine modular SLC 500 I/O addresses

INTRODUCTION

Soon you will develop your first PLC ladder diagram using a personal computer and RSLogix 500 ladder development software. Since there are many SLC 500 processors, fixed SLC 500 PLCs, and the MicroLogix PLC, lab exercises in this manual will be as generic as possible, so any of the available SLC family of PLCs can be used.

However, all exercises can be easily completed with either a MicroLogix 1000 PLC, a fixed SLC 500 PLC, or a modular SLC 500 PLC. Listed below are the I/O specifications and addressing for either the MicroLogix 1000 or the fixed SLC 500 PLC. You can easily convert the addresses for the MicroLogix 1000 or fixed SLC 500 using the information below.

IF YOU ARE USING A MICROLOGIX 1000 PLC

The MicroLogix 1000 is a small footprint, fixed I/O PLC available in two configurations: ten inputs and six outputs, and twenty inputs and twelve outputs. MicroLogix 1000 I/O is currently not expandable beyond its current I/O count.

Manuals for the MicroLogix 1000 show the following addresses on the units themselves:

16-Point MicroLogix 1000

O/0, O/1, O/2, O/3, O/4, O/5
I/0, I/1, I/2, I/3, I/4, I/5, I/6, I/7, I/8, I/9

32-Point MicroLogix 1000

O/0, O/1, O/2, O/3, O/4, O/5, O/6, O/7, O/8, O/9, O/10, O/11
I/0, I/1, I/2, I/3, I/4, I/5, I/6, I/7, I/8, I/9 I/10, I/11, I/12, I/13, I/14, I/15, I/16, I/17, I/18, I/19

As you develop user ladder programs, the correct syntax must be used whenever entering input and output addresses. If you attempt to enter incorrect address syntax, the software will give you an error message.

MICROLOGIX 1000 INPUT AND OUTPUT ADDRESSING

Figure 9-1 illustrates the input and output screw terminal and I/O identification found on a MicroLogix 1000 1761-L16BWA.

	—	AC COM	I/0	I/1	I/2	I/3	AC COM	I/4	I/5	I/6	I/7	I/8	I/9
DC OUT													
L1	L2	GR	VAC VDC	0/0	VAC VDC	0/1	VAC VDC	0/2	VAC VDC	0/3	VAC VDC	0/4	0/5

Figure 9-1   MicroLogix 1000 1761-L16BWA screw terminal identification.

As we proceed through this lesson, remember that address identification on the MicroLogix 1000 PLC is identified differently than you find in the software. The RSLogix software will display what is referred to as the formal address.

Attempting to configure any other I/O address—even using the correct format—which is not a normal, valid address for a MicroLogix 1000, will result in an "I/O address not configured" message. Keep in mind that a properly formatted I/O address which would be valid for a larger SLC 500 PLC can be programmed in a MicroLogix 1000 ladder program, even though the address may not be a valid MicroLogix 1000 address. Addressing errors will have to be corrected before the software will allow you to download and go on-line.

MICROLOGIX 1000 I/O ADDRESSING FORMAT

The SLC 500 family of PLCs includes the MicroLogix PLC's address format, which is divided into a file identifier, a slot number, a word number, and a bit number. Since the SLC 500 family of PLCs are 16-bit computers, internal data storage is in a 16-bit word format. Figure 9-2 illustrates a 16-bit word format. The I/O section is assigned a minimum of one 16-bit input and output word. Each screw terminal is assigned a number called its address. The first screw terminal is assigned the address of 0; screw terminals are assigned addresses 0 through 15. An I/O screw terminal number associates with the bit number in the 16-bit word assigned in memory to the module. Fixed PLCs, such as the MicroLogix 1000 and the older SLC 500 fixed PLC, do not have I/O modules, but each I/O screw terminal follows the same basic identification format. The top row of Figure 9-2 illustrates a 16-bit word with the bits numbered 0 through 15. Each bit number associates with the screw terminal with the same address. The bottom row contains 1s and 0s simulating the screw terminal's on or off status of the input field device wired to the I/O module's screw terminal with the same number or address. Input words are stored in the input data file, also called the input status file. Output words are stored in the output data file, which is also referred to as the output status table.

15	14	13	12	11	10	9	8	7	6	5	4	3	2	1	0
1	0	1	0	1	0	1	0	1	0	1	0	1	0	1	0

Figure 9-2   A 16-bit word representing an input or output word.

If our PLC had 24 or 32 input or output points, there would not be enough positions, or bits, in a 16-bit word to represent the I/O data from each screw terminal. Multiple words are used when there is more data that a single word can accommodate. Sixteen inputs or outputs can be represented in one 16-bit word. For more than 16 inputs or outputs, an additional word is required. Figure 9-3 illustrates two 16-bit words. Now we have up to 32 input or output data storage locations or 32 bits.

	15	14	13	12	11	10	9	8	7	6	5	4	3	2	1	0
Word 0																
Word 1																

Figure 9-3   Using two 16-bit words to accommodate 32 I/O points.

Notice that the first word is always word zero while the first screw terminal bit identifier is always bit zero. Figure 9-4 illustrates both an input and an output address and their respective parts.

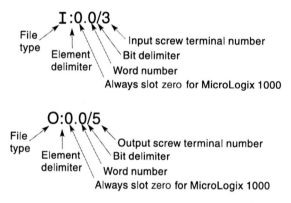

Figure 9-4   SLC 500 family input and output addressing format.

The address contains the following information:

File Type

- Inputs will be identified as I.
- Outputs will be identified as O.

Slot Number

A modular PLC has slots in its chassis that I/O modules slide into. A modular PLC consists of a PLC processor and assorted separate input and output modules. The processor for a modular SLC 500 always resides in the left-most slot in the chassis, which is identified as slot zero. I/O modules are inserted into slots beginning with slot one. Figure 9-5 identifies the parts of an SLC 500 modular chassis. Notice the slot numbers across the bottom of the figure.

Since a MicroLogix 1000 is a fixed PLC where the processor, power supply, and I/O are all built into one assembly, the processor always resides in slot zero. Therefore, the MicroLogix 1000 I/O will also be considered slot zero.

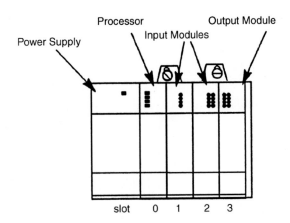

Figure 9-5   Allen-Bradley SLC 500 modular PLC. (Used with permission of Rockwell Automation, Inc.)

Word Number

The first word in memory is word zero. One word, word zero, will provide sixteen bits, one for each of the sixteen inputs or outputs. If more than sixteen I/O are available, a second word will be required. The second word is word one.

Input Screw Number

Each input or output screw terminal requires a memory location to store its on or off status. Each memory word has sixteen bits, bits 0 through 15, as illustrated in Figure 9-3. Each screw terminal associates with its bit in PLC memory.

The input or output data just introduced is stored in PLC data memory. Input data words are stored in their own file, called the input status file, or sometimes called the input status table. Likewise, the output data words have their own file. Output words are stored in the output status file or table. Figure 9-6 illustrates an RSLogix 500 software screen print with the input and output status file locations identified in the project tree. Also illustrated is the input status file showing two 16-bit input words. Word I:0.0 bits 0 through 15 associate with input screw terminals 0 through 15. Input screw terminals 16 through 23, or 31 if the I/O count of your particular MicroLogix 1000 is greater than 16 inputs, are represented in word I:0.1. Points referenced in Figure 9-6 are identified below.

- A. Data Files folder
- B. Output Status file
- C. Input Status file
- D. Display of input status data file
- E. Click on X to close file.
- F. Bit 0 of input word 0. This is address I:0.0/0.
- G. Input word 1
- H. Data File Radix. The display Radix can be changed here.
- I. Increment or decrement to view another data file.

Address Format for Input Screw Terminals Higher than Input or Output Terminal 15

If your PLC has more than 16 inputs or outputs, the I/O points above 15 go into the next word. Because of this, the address will refer to the proper bit position in the next word. As an example, if you typed in the address I:0/16 or I:0.0/16, representing screw terminal 16 in your RSLogix 500 software, the address would be displayed as I:0.1/0. This is because screw terminal 16 is the first bit position, bit 0, in word 1. Identifier (G) in Figure 9-6 is pointing to input word 1, bit 0. This is I:0.1/0, or bit 16. Refer to Figure 9-7 to view these addressing formats.

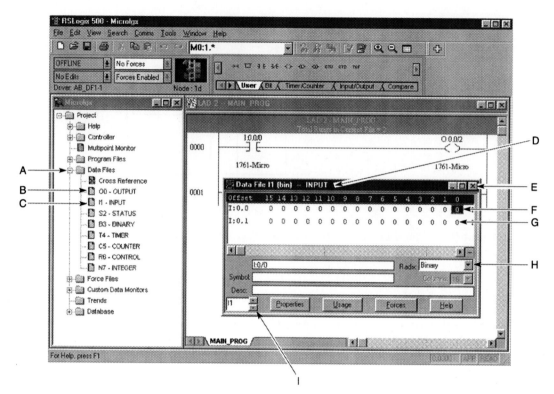

Figure 9-6   RSLogix 500 screen print with input status table file open.

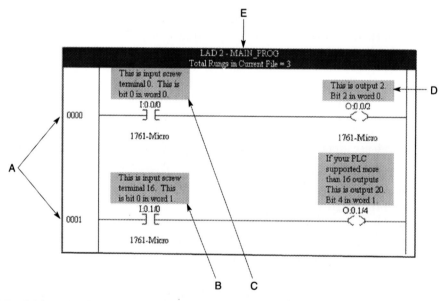

Figure 9-7   Addressing formats for SLC 500 family PLCs with more than 16 inputs (for either a fixed PLC or for a specific slot in an SLC 500 modular setup).

A. Rung numbers
B. Input I/16 is printed on MicroLogix 1000 hardware.
C. Input I/0 is printed on MicroLogix 1000 hardware identifying the screw terminal.
D. Output O/2 is printed on MicroLogix 1000 identifying this screw terminal address.
E. Identifies this as the main ladder program, ladder file 2. This file has three rungs.

Formal addresses are listed below.

The proper formal addressing syntax for the MicroLogix 1000 is as follows:

Input zero	I:0.0/0	Output zero	O:0.0/0
Input one	I:0.0/1	Output one	O:0.0/1
Input two	I:0.0/2	Output two	O:0.0/2
Input three	I:0.0/3	Output three	O:0.0/3
Input four	I:0.0/4	Output four	O:0.0/4
Input five	I:0.0/5	Output five	O:0.0/5
Input six	I:0.0/6	Output six	O:0.0/6
Input seven	I:0.0/7	Output seven	O:0.0/7
Input eight	I:0.0/8	Output eight	O:0.0/8
Input nine	I:0.0/9	Output nine	O:0.0/9
Input ten	I:0.0/10	Output ten	O:0.0/10
Input eleven	I:0.0/11	Output eleven	O:0.0/11
Input twelve	I:0.0/12		
Input thirteen	I:0.0/13		
Input fourteen	I:0.0/14		
Input fifteen	I:0.0/15		
Input sixteen	I:0.1/0		
Input seventeen	I:0.1/1		
Input eighteen	I:0.1/2		
Input nineteen	I:0.1/3		

Figure 9-8 lists current MicroLogix 1000 catalog numbers and the available number and type of I/O.

MICROLOGIX 1000 PROGRAMMABLE CONTROLLERS			
Catalog Number	**Line Power**	**Inputs**	**Outputs**
1761-L10BWA	120/240V ac	(6) DC Sink/Source	(4) Relay
1761-L10BWB	24V dc	(6) DC Sink/Source	(4) Relay
1761-L16AWA	120/240V ac	(10) 120V ac	(6) Relay
1761-L16BWA	120/240V ac	(10) DC Sink/Source	(6) Relay
1761-L16BWA	24V dc	(10) DC Sink/Source	(6) Relay
1761-L16BBB	24V dc	(10) DC Sink/Source	(4) FET, (2) Relay
1761-L20AWA-5A	120/240V ac	(12) 120V ac, (4) Analog	(8) Relay, (1) Analog
1761-I20BWA-5A	120/240V ac	(12) DC Sink/Source, (4) Analog	(8) Relay, (1) Analog
1761-L20BWB-5A	24V dc	(12) DC Sink/Source, (4) Analog	(8) Relay, (1) Analog
1761-L32AAA	120/240V ac	(20) 120V ac	(10) Triac, (2) Relay
1761-L32AWA	120/240V ac	(20) 120V ac	(12) Relay
1761-L23BWA	120/240V ac	(20) DC Sink/Source	(12) Relay
1761-L32BWB	24V dc	(20) DC Sink/Source	(12) Relay
1761-L32BBB	24V dc	(20) DC Sink/Source	(10) FET, (2) Relay

Figure 9-8   MicroLogix 1000 Programmable Controllers Catalog Number Identification. (Data compiled from Allen-Bradley technical documentation).

IF YOU ARE USING A FIXED SLC 500 PLC

The fixed SLC 500 was the original fixed member of the SLC family, before the development of the MicroLogix 1000. The fixed SLC contains the processor, power supply, and I/O built into a single unit, just like the MicroLogix 1000. All I/O is built in, or fixed. The number of, and signal type of, I/O in a fixed PLC is not changeable. If a fixed PLC was ordered with twenty 120 VAC input points, that voltage level and number of inputs are fixed at the factory. The end user cannot change the I/O type or count.

SLC 500 fixed I/O PLC controllers come in three I/O configurations: 20, 30, or 40 I/O points. One optional two-slot expansion chassis can be clipped to the right end of the unit to add up to two additional *modular* I/O modules.

Using the optional two-slot expansion slot chassis, two I/O modules may be inserted in the expansion chassis, increasing the possible I/O count by 64 I/O. The expansion chassis can include input or output modules up to thirty-two points each. Specialty modules, like analog input or output, may reside in the expansion chassis.

SLC 500 fixed PLCs are available as listed below (note that the last part of the part number indicates the total of that PLC's inputs and outputs).

1747-L20

Twelve inputs, addresses: I:0.0/0 through I:0.0/11.
Eight outputs, addresses: O:0.0/0 through O:0.0/7.

1747-L30

Eighteen inputs, addresses: O:0.0/0 through I:0.0/1.
Twelve outputs, addresses: O:0.0/0 through O:0.0/11.

1747-L40

Twenty-four inputs, addresses: I:0.0/0 through O:0.1/8.
Sixteen outputs, addresses: I:0.0/0 through O:0.0/15.

The table in Figure 9-9 identifies the letter and the corresponding I/O configuration:

Part Number	Inputs	Outputs	Line Voltage
1747-L_A	120 VAC	Relay	120/240 VAC
1747-L_B	120 VAC	Triac	120/240 VAC
1747-L_C	DC Sink	Relay	120/240 VAC
1747-L_D	DC Sink	Triac	120/240 VAC
1747-L_E	DC Sink	DC Source	120/240 VAC
1747-L_F	DC Sink	Relay	24 VDC
1747-L_G	DC Sink	DC Source	24 VDC
1747-L_L	DC Source	DC Sink	120/240 VAC
1747-L_N	DC Source	DC Sink	24 VDC
1747-L_P	240 VAC	Triac	120/240 VAC
1747-L_R	240 VAC	Relay	120/240 VAC

Figure 9-9   SLC 500 fixed PLC part numbers and I/O correlation.

I/O Programming Addresses

As you develop user ladder programs, the proper syntax must be used whenever entering input and output addresses. If you attempt to enter incorrect address syntax, the software will give you the same error message as the MicroLogix 1000. Addressing a fixed SLC 500 also follows the same format as the MicroLogix 1000.

IF YOU ARE USING A MODULAR SLC 500 PLC

When complete flexibility in I/O count and I/O mix, processor power features, and memory size is desired, a modular PLC is required for the application. A modular SLC 500 can be specifically configured to fit almost any application. Being modular, the processor, power supply, chassis, and required I/O modules are individually selected from the many selections available.

Chassis are used to hold all the selected pieces together. Chassis are selected by the number of *slots* available for module insertion. Typically, any module can be inserted into any slot. Slots are numbered from left to right starting at slot zero. The processor always goes in slot zero. Even though the processor is in slot zero, a modular PLC, unlike a fixed PLC, has no I/O associated with slot zero. All chassis start with slot zero as the left-most slot in the chassis. SLC 500 chassis slot numbers are listed below:

> A four-slot chassis has slots zero through three.
> A seven-slot chassis contains slots zero through six.
> The ten-slot chassis consists of slots zero through nine.
> Thirteen-slot chassis have slots zero through twelve.

Figure 9-10 illustrates SLC chassis and slot assignments.

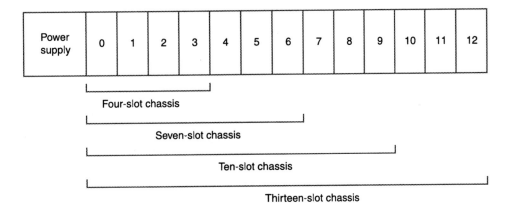

Figure 9-10   SLC 500 chassis slot numbering assignments.

The modular SLC 500 PLC may contain up to three chassis connected together. Any mix of the available chassis may be connected to increase the total I/O count; a total of thirty I/O slots is allowed in any modular SLC 500 PLC. Figure 9-11 is an example of a three-chassis SLC 500 PLC.

Determining a Discrete I/O Address

The processor must be able to differentiate one I/O module from another in each chassis. Each I/O point on each module also must be uniquely identified.

In order to uniquely identify a module type, slot, and I/O point, an *address* is constructed using all of this information. The module type, O for output and I for input, the chassis slot number, and the module's individual I/O screw terminal data are used to construct the input or output point address. This I/O address is the identifying data you will associate with each instruction on a ladder program.

Each chassis slot may be either an input or an output. If using combination input and output modules, a single slot can contain input as well as output addresses. Each I/O point must have a

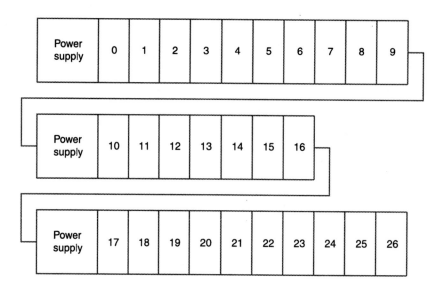

Figure 9-11 Three SLC 500 chassis expanded locally to a total of twenty-seven slots.

unique address which is reflected in either the input data file or the output data file. The addressing format is very similar to the MicroLogix or fixed SLC 500 PLCs. The only difference between fixed I/O addressing and modular I/O addressing is the slot number will always be zero for fixed PLCs whereas the slot number for addressing I/O points for a modular SLC 500 PLC will be any valid slot value from one to thirty—but never zero. The addressing format for outputs consists of module type (O for output), chassis slot number (one to thirty), and the module's individual output screw terminal (0 to 31). Figure 9-12 illustrates addressing format.

The addressing format for inputs is very similar to output addressing. Figure 9-13 illustrates input addressing format.

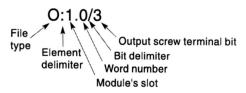

Figure 9-12 Modular SLC 500 output addressing format.

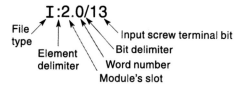

Figure 9-13 Modular SLC 500 input addressing format.

When 32-point I/O modules are used in a modular SLC 500 PLC, two 16-bit data words are necessary to store the 32 input or output bits. The addressing format is the same as for fixed PLCs.

I/O Address Conversion for Your Classroom SLC 500 PLC Trainer

These lab manual exercises were developed using an SLC 500 seven-slot modular chassis with a 5/04 processor. If you are using a fixed PLC, such as a MicroLogix 1000 or an SLC 500 fixed PLC, address conversion needs to be completed as you enter ladder logic for future lab exercises. Address conversion is easy. Follow these steps to convert the modular I/O addresses:

1. Remember that a fixed PLC, such as the MicroLogix 1000 and the SLC 500 fixed unit without I/O expansion, is always slot zero.
2. Understand which PLC you are working with.
3. Know how many input and output points your PLC has.

4. Start with the modular I/O address and change the slot number to zero.
5. Verify that your fixed PLC has its inputs and outputs wired as you expect so your demo programs operate correctly.

Figure 9-14 can be used to convert modular PLC input and output addresses used for exercises in this manual to either fixed SLC 500 addresses or MicroLogix addresses. Notice that the slot addresses for the modular addresses currently contain an "x." The x signifies any valid slot address for a modular PLC.

CONVERTING MODULAR I/O ADDRESSING TO FIXED I/O ADDRESSING			
Modular Input Address	Fixed Input Address	Modular Output Address	Fixed Output Address
I:x.0/0	I:0.0/0	O:x.0/0	O:0.0/0
I:x.0/1	I:0.0/1	O:x.0/1	O:0.0/1
I:x.0/2	I:0.0/2	O:x.0/2	O:0.0/2
I:x.0/3	I:0.0/3	O:x.0/3	O:0.0/3
I:x.0/4	I:0.0/4	O:x.0/4	O:0.0/4
I:x.0/5	I:0.0/5	O:x.0/5	O:0.0/5
I:x.0/6	I:0.0/6	O:x.0/6	O:0.0/6
I:x.0/7	I:0.0/7	O:x.0/7	O:0.0/7
I:x.0/8	I:0.0/8	O:x.0/8	O:0.0/8
I:x.0/9	I:0.0/9	O:x.0/9	O:0.0/9
I:x.0/10	I:0.0/10	O:x.0/10	O:0.0/10
I:x.0/11	I:0.0/11	O:x.0/11	O:0.0/11
I:x.0/12	I:0.0/12	O:x.0/12	O:0.0/12
I:x.0/13	I:0.0/13	O:x.0/13	O:0.0/13
I:x.0/14	I:0.0/14	O:x.0/14	O:0.0/14
I:x.0/15	I:0.0/15	O:x.0/15	O:0.0/15

Figure 9-14   Conversion table from modular SLC 500 to fixed MicroLogix 1000 of fixed SLC 500.

Programming I/O Addresses

Now that you have an understanding of how the RSLogix 500 software will display your I/O addresses, let's look at your options when typing I/O addresses as you develop your lab exercises and future programs. The formal input address I:0.0/1 can also be typed into your ladder program as I:0.0/1 or I:0/1. The software will automatically convert, if necessary. The table in Figure 9-15 provides examples of fixed and modular I/O addressing programming options.

It is not necessary to enter the word number when programming fixed or modular PLCs as long as the address does not contain more than sixteen bits.

RSLogix Address Wizard

A new option regarding address entry with the RSLogix 500 software is the Address Wizard. The Address Wizard was first available with release of Version 4.0 of the RSLogix software. The Wizard will assist in address selection by selecting the next available address. With a couple of mouse clicks you can insert it for the instruction currently being programmed. The Wizard can be used to automatically select the next available address or to allow manual address entry, or it can be disabled if not used. You will be introduced to and begin to work with the Address Wizard after we start developing ladder programs.

Address Printed on PLC	Type Address in Software as	Software Displayed on Ladder Rung
I/0	I:0/1 or I:0.0/1	I:0.0/1
I/12	I:0/12 or I:0.0/12	I:0.0/12
I/15	I:0/15 or I:0.0/15	I:0.0/15
I/16	I:0/16 or I:0.1/0	I:0.1/0
I/17	I:0/17 or I:0.1/1	I:0.1/1
I/19	I:0/19 or I:0.1/3	I:0.1/3
I/23	I:0/23 or I:0.1/7	I:0.1/7
I:3.0/15	I:3/15 or I:3.0/15	I:3.0/15
I:3.0/23	I:3/23 or I:3.1/7	I:3.1/7
O:4.0/19	O:4/19 or O:4.1/3	O:4.1/3

Figure 9-15   RSLogix 500 I/O address entry options.

ADDRESSING EXERCISES

1. Complete the address format information in Figure 9-16.

Input or Output?	Chassis Slot?	Screw Terminal?	Address
I	1	4	I:1/4
Output	2	?	O:2/13
Input	?	1	I:4/1
Input	12	11	?
?	11	?	O:11/9
Output	3	3	?
Input	?	?	I:14/5
Input	9	15	?
Input	1	?	I:1/8
?	?	?	O:13/15
Output	5	?	O:5/9
Input	6	0	?

Figure 9-16   SLC 500 addressing exercise.

2. Illustrated in Figure 9-17 is a seven-slot SLC 500 chassis. Fill in the appropriate bit positions in the associated output file (Figure 9-18) and input file (Figure 9-19) with the appropriate status bits for each module. The module in slot one, an input module, has been completed as a guide.

Power supply	Slot 0	Slot 1	Slot 2	Slot 3	Slot 4	Slot 5	Slot 6
	CPU	Input	Input	Output	Input	Output	Output

Figure 9-17   Modular SLC 500 representation.

The module in slot one has screw terminals 0, 1, 7, 9, and 14 as ON conditions.
The module in slot two has screw terminals 2, 3, 5, 9, and 10 as ON conditions.
The module in slot three has screw terminals 0, 1, 3, and 15 as ON conditions.
The module in slot four has screw terminals 2, 7, 8, 9, and 14 as ON conditions.
The module in slot five has screw terminals 1, 7, 9, 11, and 14 as ON conditions.
The module in slot six has screw terminals 14 and 15 as ON conditions.

	15	14	13	12	11	10	9	8	7	6	5	4	3	2	1	0
0:3.0																
0:5.0																
0:6.0																

Figure 9-18 Output status file for the PLC represented in Figure 9-17.

	15	14	13	12	11	10	9	8	7	6	5	4	3	2	1	0
I:1.0	0	1	0	0	0	0	1	0	1	0	0	0	0	0	1	1
I:2.0																
I:4.0																

Figure 9-19 Input status file for the PLC represented in Figure 9-17.

3. Determine addresses of the following module in chassis slot four, Figure 9-20.

Is this an input or output module? _____

PB1 address is _____

LS3 address is _____

LS4 address is _____

SW2A address is _____

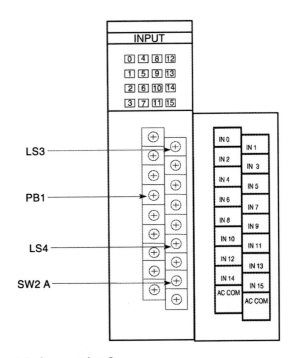

Figure 9-20 I/O module for question 3.

4. Determine addresses of the following module in chassis slot fifteen, Figure 9-21.

Is this an input or output module? _____

A address is _____

B address is _____

C address is _____

D address is _____

E address is _____

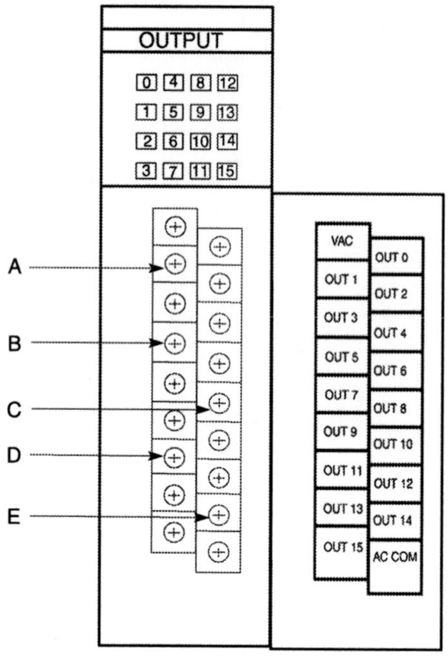

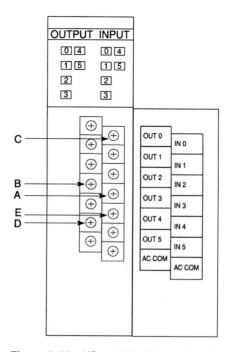

Figure 9-21 I/O module for question 4.

5. Determine addresses of the following module in chassis slot three, Figure 9-22.

Is this an input or output module? _____

B address is _____

A address is _____

_____ is connected to address I:3/0

_____ is connected to address O:3/5

E is _____

Figure 9-22 I/O module for question 5.

REVIEW QUESTIONS

1. Fixed PLCs are considered to be in which slot regarding I/O addressing? _____

2. SLC 500 modular chassis come in what sizes? _____

3. A modular SLC 500 processor always goes in which chassis slot? _____

4. Three thirteen-slot SLC 500 modular I/O chassis are connected together during installation of the system. The processor is placed in slot zero. I/O modules are placed in the remaining slots.

   Describe any problems with this configuration. _____

   _____

   _____

5. Consider the following address: O:0.1/1.

   A. In what type of PLC would you find this? _____

   B. What screw terminal would you find the field device wired to? _____

   C. How would the screw terminal be physically identified on the PLC? _____

   _____

   D. Where would this input bit be found in the RSLogix 500 data files? _____

6. What is the input point address of screw terminal 4 on an input module in chassis slot fourteen?

   _____

7. Explain how the input data file associates with a screw terminal on an input module.

   _____

   _____

8. The SLC 500 PLCs are 16-bit computers. Explain how this correlates to the input or output data file. _____

   _____

9. A 32-point output module resides in slot twenty-two of the SLC chassis. What is the address of screw terminal 24? _____

10. Computers always identify the first bit, first file, or first word as word _____.

11. What is the total number of chassis that can be connected together in a local I/O SLC 500 configuration? _____

12. How many valid slots can be configured in a modular SLC 500 system? _____

13. The power supply is attached on the far left side of a modular chassis. Since the power supply is part of slot zero in a fixed PLC, does the power supply occupy slot zero in a modular configuration? _____

14. Identify each piece of the following address:

    I : 0 . 0 / 9

15. Consider the following address: I:0.1/2.

    A. In what type of PLC would this input point be found? _____

    B. What screw terminal would you find the field device wired to? _____

    C. How would the screw terminal be physically identified on the PLC? _____

    _____

    _____

    D. Where would this input bit be found in the RSLogix 500 data files? _____

16. Consider the following address: I:9.1/3.

    A. In what type of PLC would this input point be found? _____

    B. What screw terminal would you find the field device wired to? _____

    C. Identify the physical hardware location of this I/O point. _____

    _____

    D. Where would this input bit be found in the RSLogix 500 data files? _____

17. The address O/5 is found on the PLC above the screw terminal.

    A. What is the formal address of this I/O point? _____

18. Refer to Figure 9-23. Identify each part of the address.

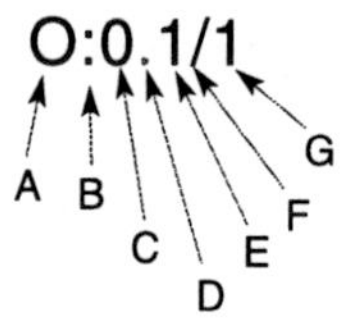

O:0.1/1
A B C D E F G

Figure 9-23 Address format identification.

A. _____

B. _____

C. _____

D. _____

E. _____

F. _____

G. _____

10

Creating a New Project

OBJECTIVES

Upon completion of this laboratory exercise, you should be able to:

- create a new RSLogix 500 project file
- read a processor's I/O configuration
- configure a processor's communication channels
- configure the ladder window properties
- save the project

INTRODUCTION

Each time a new SLC 500 or MicroLogix PLC ladder program is developed, a new project file must be created. The project file will be created using RSLogix 500 software. Included in the creation of the processor file is a name for the file, data entered on the processor and the processor operation system, and chassis and configuration of the I/O modules in the chassis(s).

This lab exercise will guide you through the procedure to create a new processor file, enter processor and operating system data, configure the PLC's I/O, name the project file Begin, and save the newly created processor file to your floppy disk. This floppy disk will be your beginning, or default, project file for future programming exercises. Using this as a default file, you will not have to go through all the steps to create a new file each time you begin a new programming exercise.

DEVELOPING A NEW PROJECT FILE

You will always begin at the main window. The main window is the first screen that appears when you start up the RSLogix software. The main window shows the RSLogix 500 toolbars. Refer to Figure 10-1. From this window there are four options:

1. *Create a new project*. Click on the Open New File icon (A) on the Windows toolbar to start a new project.

2. *Open an existing project*. Open an existing project (B) that is currently stored on a hard drive or other disk drive. After the project is open, the project can be modified or downloaded to the processor. At this point there is no communication with the processor. In other words, you are not on-line.

3. *Upload*. To upload is to take a copy of the processor file residing in the PLC processor and copy it to the personal computer's hard drive. Refer to (C) in the figure.

4. *Go on-line*. Go on-line with a processor to monitor or modify the existing processor file while it is operational. Refer to (D) in the figure. Note that currently the software is off-line (E). Figure 10-1 illustrates these options on the RSLogix main window.

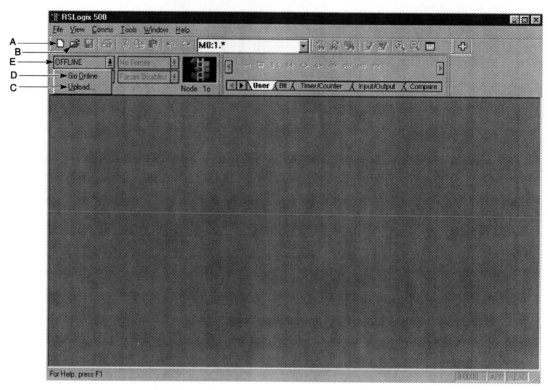

Figure 10-1 RSLogix 500 main window.

For this exercise we will create a new project which we will call Begin. We will go through the steps to create and configure our new project so we can begin to enter ladder logic. When completed, we will enter a revision note and save the project to a floppy disk for future use.

LAB ONE

1. _____ From the RSLogix 500 main window, click on the Open New File icon.
2. _____ The Select Processor Type dialog box will open and look similar to Figure 10-2.
3. _____ Enter the processor name of Begin. Refer to (A) as illustrated in Figure 10-2.
4. _____ The SLC family processor should have been recorded from an earlier lab exercise. Select your processor out of the list. Make certain you select the proper processor, operating system, and series. Refer to (B).
5. _____ Verify communications settings by selecting the correct driver (C) and node address (D) for the processor used in this project.
6. _____ If you do not know the correct driver and node address, click on Who Active (E). This will take you to RSLinx and the Who Active screen. Here you can select the proper driver and processor node address.
7. _____ Click on the Help button (F) to view the Help screens. Help provides information on each field of the dialog box.
8. _____ When completed with Help, select file and close to return to the Select Processor Type dialog box.
9. _____ When you are finished working in this dialog box, click on OK (G).
10. _____ The computer will create all project files as it works and adds the project tree and ladder window to the main window. Refer to Figure 10-3 for important information displayed in the window.

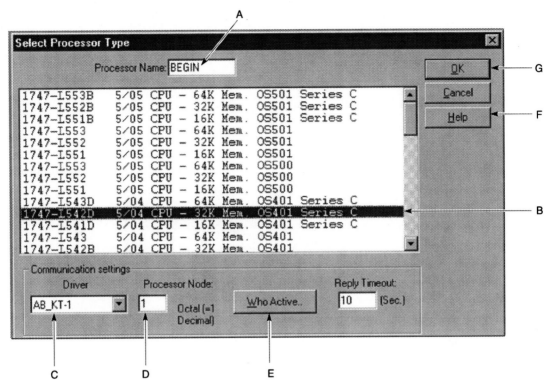

Figure 10-2   Select Processor Type.

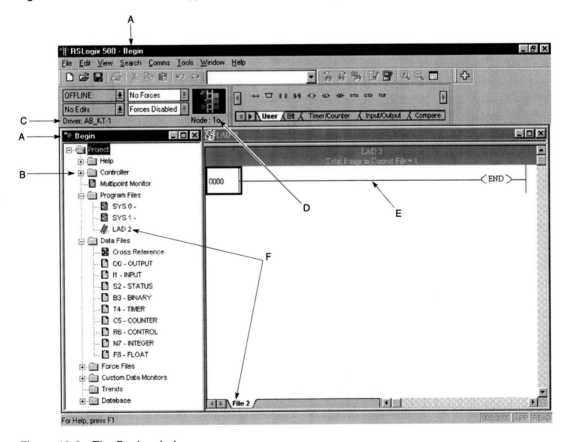

Figure 10-3   The Begin window.

A. Project name
B. Controller folder
C. RSLinx drive name selected
D. RSLinx node address selected
E. Ladder rung 0000
F. Identifies this as main ladder file 2

11. _____ The processor communication channels may need to be configured if other than default communication parameters are needed. Click on the + sign in front of the Controller folder (B) in Figure 10-3 to expand the folder. Double-click on Channel Configuration in the project tree to view the Channel Configuration windows illustrated in Figure 10-4 (A).

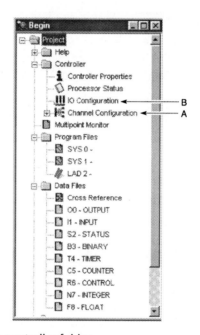

Figure 10-4 Project Tree controller folder.

Figure 10-5 illustrates the Channel Configuration tabs for a 5/04 modular processor. The 5/04 processor has two communication channels. Channel 0 is the 9-pin D-shell serial connection. Channel 1 is for a data highway plus (DH+) network connection. The General tab provides an overview of the Channel 0 and Channel 1 communication setups.

The default communication setup for channel 0 is Channel 0 System, which supports a serial communication link between a personal computer and the processor for programming. Channel 1 default is Data Highway Plus. The other option for channel 1 is shutdown. Under normal conditions channel configuration defaults are acceptable.

12. _____ Click OK to close the Channel Configuration window. Next, I/O configuration will be completed.

If you have other than a 5/03 or above modular processor, skip to "Configuring a Fixed SLC 500 PLC or MicroLogix 1000."

If you have a fixed PLC, skip to step 20.

13. _____ Double-click on I/O configuration (B) as illustrated in Figure 10-4. If you are using a 5/03 or higher modular processor, simply click on the Read IO Config button (A) in Figure 10-6.

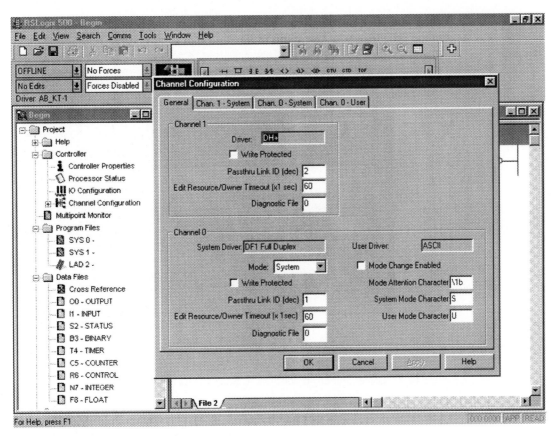

Figure 10-5   Channel Configuration tabs.

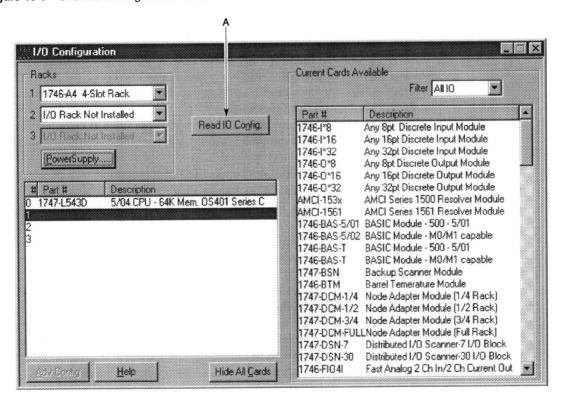

Figure 10-6   Automatic Read I/O Configuration for a 5/03 and above modular processor.

14. _____ When you click on the Read IO Config button, the Read IO Configuration from Online processor window opens as illustrated in Figure 10-7.

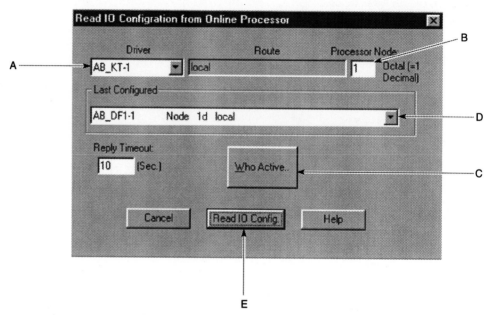

Figure 10-7 Driver selection window to Read I/O Configuration.

15. _____ Verify that the correct RSLinx driver is selected. To change the driver, refer to Figure 10-7 along with the window options listed below.
   A. Click on arrow to open Driver selection drop-down box. Click on proper RSLinx driver.
   B. Click in box to type in correct processor node.
   C. If you are not sure of the proper processor node address, click on the Who Active button. This will display the Who Active screen from RSLinx.
   D. Another option is to click on the arrow and see if the driver you need is available in the last configured drop-down list.
   E. When the proper driver and node are selected, press Read IO Config. The rack section (A) in Figure 10-8 should be correct. Also the chassis configuration should be displayed as illustrated in (B).

16. _____ If you have a processor that does not support I/O configuration:
   1. Manually select your rack(s) (A) in Figure 10-8.
   2. Modules displayed can be filtered by clicking in the arrow (C) and selecting the modules to be displayed.
   3. Manually select each I/O module from the list (D) in Figure 10-8. Select the module, hold down the left mouse button, and drag the module to the proper slot position (B).

17. _____ Click on Power Supply button (E) Figure 10-8. The Power Supply Loading window should open as illustrated in Figure 10-9. This feature of the RSLogix software is informational only, that is, the calculations are completed for you using information you select in this window in conjunction with the I/O configuration modules and slot information. Notice the five sections of the window:

   Rack. Select the rack you wish to determine power supply loading for (A). Keep in mind that information from the I/O configuration will be used to calculate power supply loading.

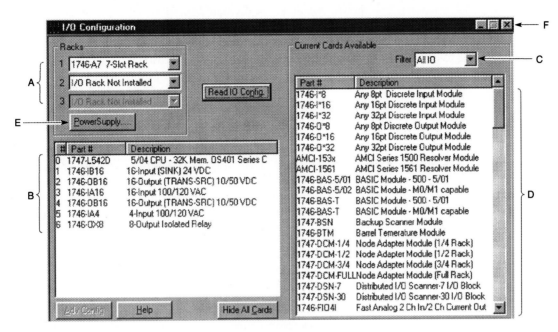

Figure 10-8   Completed Read I/O Configuration.

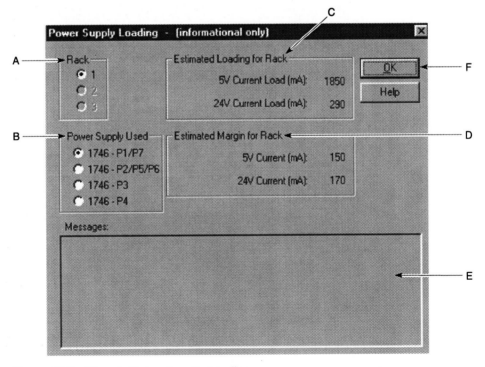

Figure 10-9   Chassis Power Supply Loading.

Power Supply Used. Select the power supply for the chassis you wish (B) power supply loading calculated for.

Estimated Loading for Rack. The calculated loading for the rack and power supply selected is displayed in (C).

Estimated Margin for Rack. The estimated margin, or shortfall for the rack power supply loading is displayed in (D).

Messages. If the power supply loading calculation results in an overloaded power supply, a message identifying an overloaded power supply will be displayed at (E).

18. _____ Click OK (F) to close the Power Supply Loading window.
19. _____ Close the I/O Configuration window. Refer to (F) in Figure 10-8.

CHANNEL SETUP AND DOWNLOADING TO AN SLC 5/05 ETHERNET PROCESSOR

Before starting the lab, have your instructor walk you through the steps of setting up the IP address for your personal computer. For this exercise, use the IP address 100.100.100.3.

You configured an Ethernet RSLinx driver in Lab Exercise 5. Before you can communicate with or view the processor on the RSWho screen, you need to set up the Ethernet IP address for the SLC 5/05 processor channel configuration.

1. _____ Open the RSLogix 500 software.
2. _____ Create a new project.
3. _____ Fill in a name for your processor. Refer to Figure 10-10.

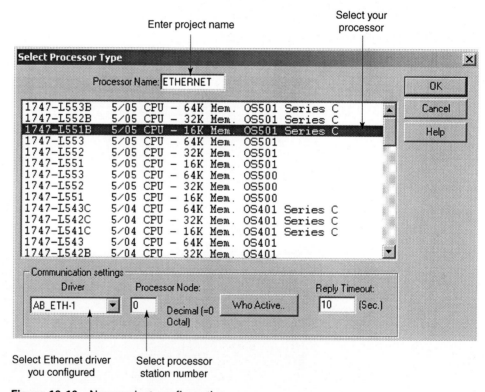

Figure 10-10   New project configuration.

4. _____ In the Select Processor Type dialog box, select the SLC 5/05 processor that you are using.
5. _____ Select the Ethernet driver, AB_ETH-1, from the drop-down list under Communication Settings and Driver. Refer to Figure 10-10 if necessary.
6. _____ Select Processor Node as 0. (In Lab Exercise 5, you configured an RSLinx Ethernet Driver. You entered an IP address and noted the resulting station number. Refer to Figure 5-44 and Lab step 9.)
7. _____ Click OK when completed.

8. _____ Open the Controller folder and double-click on Channel Configuration. See Figure 10-11.

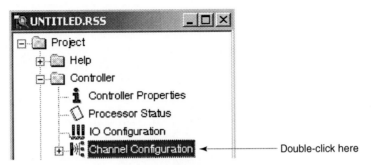

Figure 10-11   Double-click for Channel Configuration.

9. _____ Click on the Channel 1 System tab as shown in Figure 10-12. Channel 1 tab opens to display Channel information. See Figure 10-13.

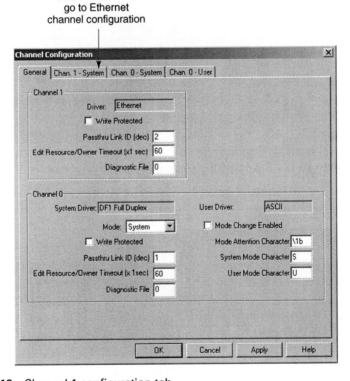

Figure 10-12   Channel 1 configuration tab.

10. _____ Enter the IP address for the SLC 5/05 processor. For this lab exercise, you will use the IP address you programmed into the RSLinx driver. Type the value illustrated in Figure 10-14. Make sure Bootp Enable is not checked.

11. _____ Next, enter the Subnet Mask shown in Figure 10-14. The Subnet Mask is used to allow or deny access to a sub network (Subnet).

12. _____ Click OK. This completes the channel configuration for this lab exercise.

The Ethernet channel configuration has now been set up in the RSLogix 500 software. Before the processor can know its IP address, the project with the new channel configuration

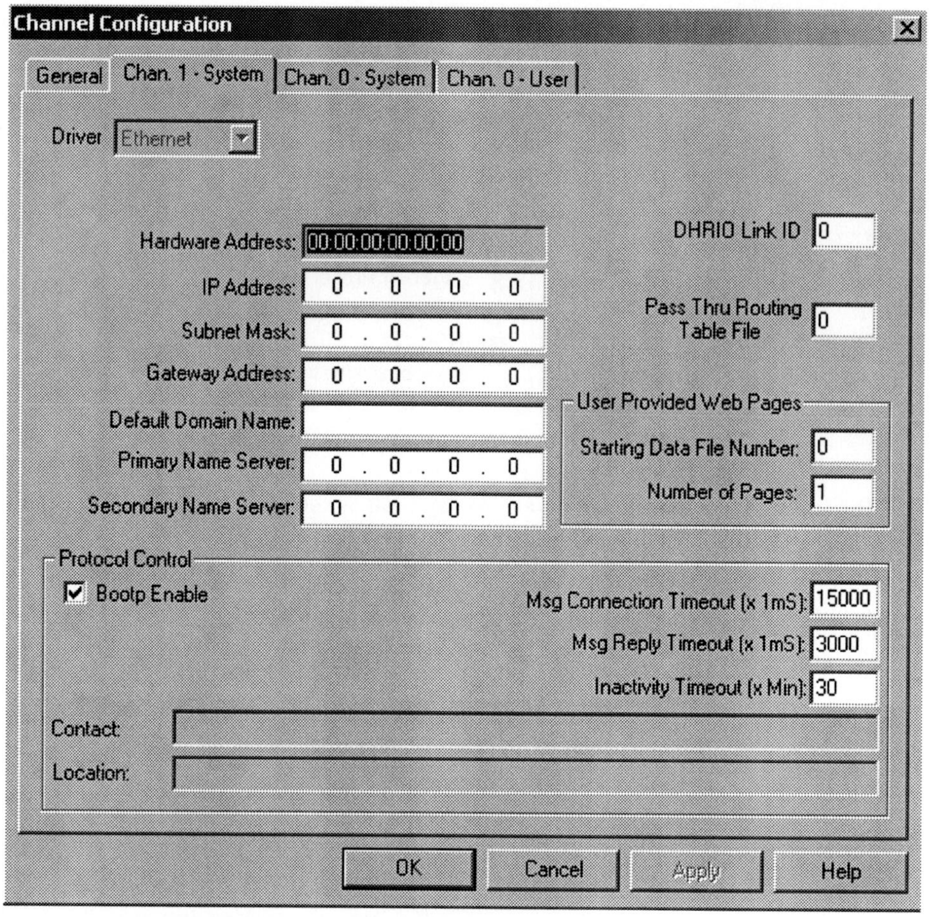

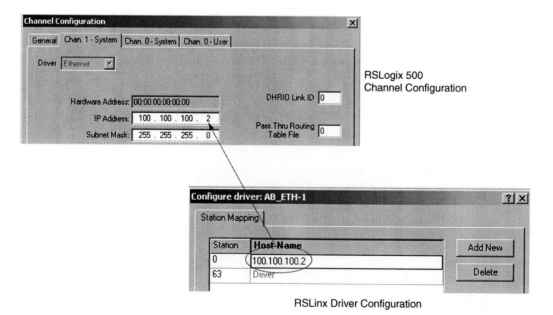

Figure 10-13   RSLogix 500 default Ethernet configuration.

Figure 10-14   Correlation between RSLogix channel configuration setup and RSLinx driver IP address.

must be downloaded to our processor. Because we cannot use Ethernet to communicate with our processor until our processor knows its Ethernet IP address, we will use the serial driver to download the new configuration to the processor.

13. _____ Download the modified project via serial cable to the SLC 500 processor.

14. _____ Since we do not know what Ethernet IP address is currently configured in the SLC 5/05 processor, we will replace, or overwrite, it with our new IP of 100.100.100.2. As part of the download, you will see a message similar to that shown in Figure 10-15. Click on Apply to overwrite the current IP address in the SLC processor. If you click on Don't Apply, the new channel IP address will not be stored in the SLC 5/05 processor. Note that the window states that there may be a loss of communication on the current (serial) channel 0. If the serial configuration was modified in the channel configuration you just completed, the processor channel will be modified to reflect the new serial configuration. In that case, the serial RSLinx driver will have to be reconfigured to reestablish communications.

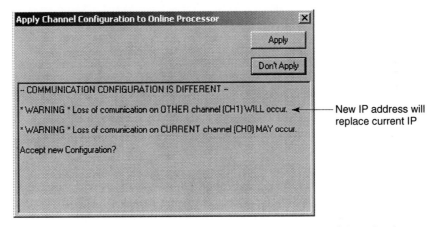

Figure 10-15   Apply channel configuration to processor as part of download.

15. _____ Now that you have downloaded the new channel configuration to the processor, you can go to RSLinx and find the processor on Ethernet on our RSWho screen. See Figure 10-16.

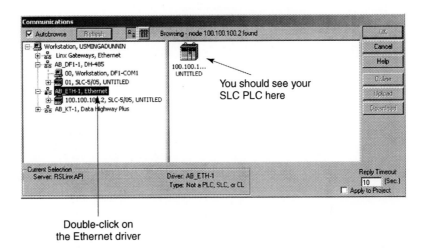

Figure 10-16   RSWho screen showing that the SLC 5/05 processor is communicating.

If the processor is not communicating, there will be an X superimposed over a question mark (?) where the processor icon would usually be. The text below the icon will read "Unrecognized Device" as illustrated in Figure 10-17. If you see this, go back and verify that the RSLinx driver, your personal computer, and the RSLogix 500 channel configuration are all set up correctly. Ask your instructor for assistance if you continue to have trouble.

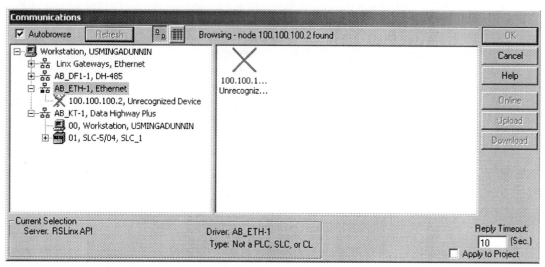

Figure 10-17 RSWho screen showing the device as unrecognized.

Now that you have established communications with your SLC 5/05 processor, let's try to download your RSLogix 500 project using Ethernet. The driver and node address on your Program/Processor Status toolbar must show the correct driver and processor station or node address. This is illustrated in Figure 10-18. The figure shows RSLinx and our configured driver. It also shows how that information must correspond with the driver and station or node address selected on your RSLogix 500 Program/Processor Status toolbar. If this information does not match, you will not be able to download to the correct processor.

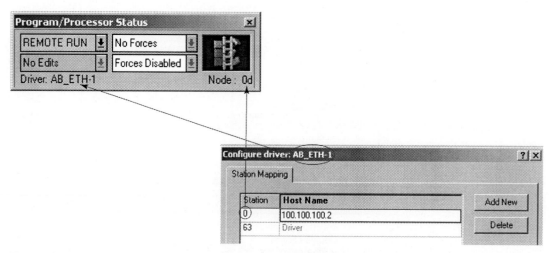

Figure 10-18 Correlation between RSLinx driver configuration and RSLogix 500 Program/Processor Status toolbar.

16. _____ To set up the correct driver information on the toolbar, click on Comms as illustrated in Figure 10-19. Then click on System Comms.

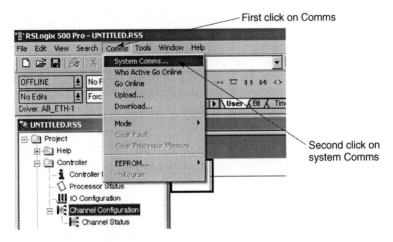

Figure 10-19   Setting up System Communications.

17. _____ A communications screen similar to that shown in Figure 10-20 should display. Select your Ethernet driver.

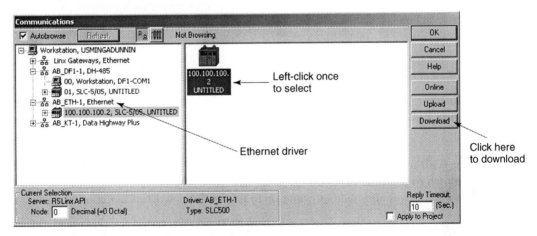

Figure 10-20   RSLogix 500 Communications screen.

18. _____ Left-click once on the SLC 500 PLC icon.
19. _____ Click Download.
20. _____ A similar message to that shown in Figure 10-21 will come up. After you verify that you are downloading to the correct processor, click Yes to continue.
21. _____ Notice that in Figure 10-22 the cautionary message states that the current project settings do not match system settings. The screen shown in the lower picture of the figure will display as part of your download. This screen shows that the current communication settings in your RSLogix 500 project do not match the settings you wish to use for the download. The upper-left picture of Figure 10-22 shows the on-line toolbar from RSLogix software. Note that the driver name is AB_DF1-1 while the node address is Node 1d. This is your serial driver.

You are now going to use the Ethernet driver to download your project. You set up this driver in steps 16 through 19 above. The message in the lower picture of Figure 10-22 asks you if

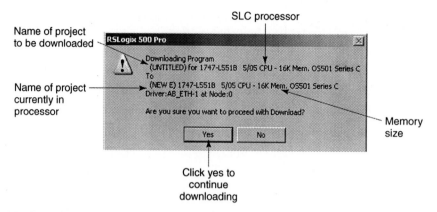

Figure 10-21   Downloading cautionary information for an SLC 500 processor.

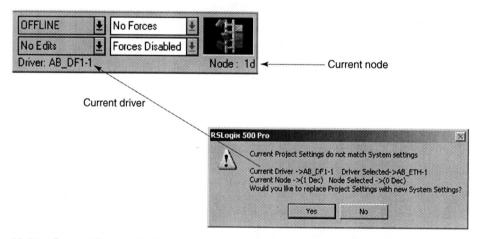

Figure 10-22   Current Project Settings do not match System settings cautionary screen.

you wish to replace the current (serial) settings with the new Ethernet settings. Click Yes as you do wish to continue to download using Ethernet. The on-line toolbar in your project should update the driver and node information to reflect the new Ethernet configuration.

22. _____ If your processor was in Run Mode as you started the download, the screen shown in Figure 10-23 may come up. The processor must be switched to Remote Program (PROG) Mode before downloading the new project to it. DO NOT move the key switch on the processor; use your mouse to click yes, and you will remotely change the operating mode to Program. The download should continue.

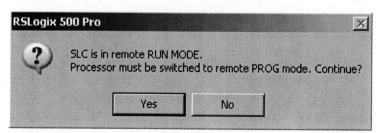

Figure 10-23   SLC processor must be switched to remote program (PROG) mode to continue download.

23. _____ When the download is complete, the message shown in Figure 10-24 will ask if you want to put the processor back into Run Mode. Click Yes.

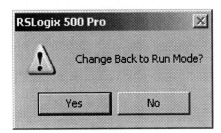

Figure 10-24   Change processor back to Run Mode?

24. _____ Refer to Figure 10-25. With the processor running, do you want to go Online? Going Online will allow you to monitor the ladder rungs as they execute. You will also be able to monitor data table values. Click Yes.

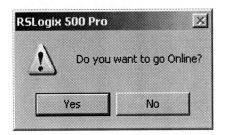

Figure 10-25   Do you want to go Online?

25. _____ If you look at the Program/Processor status toolbar, you should see that the processor is in Remote Run Mode. Current communication settings are illustrated in Figure 10-26.

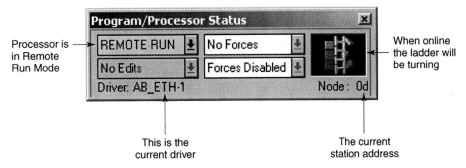

Figure 10-26   Program/Processor Status toolbar.

26. _____ Your program is running and you are on-line. You can now monitor your ladder rungs and data tables.

Configuring a Fixed SLC 500 PLC or MicroLogix 1000

Since fixed PLCs have their I/O built into the same package as the processor and power supply, there is no modular I/O to configure. Fixed members of the SLC 500 family cannot auto-configure themselves. These PLCs will have to be selected from the Select Processor Type window, as illustrated in Figure 10-27. There will be no need to perform an I/O configuration. Channel configuration can be opened to modify the serial communication baud rate between the personal computer and the fixed MicroLogix PLC or the DH485 baud rate and node address.

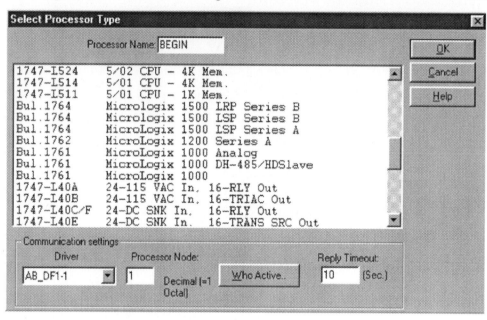

Figure 10-27 Select Processor Type window.

1. _____ Name your processor Begin.
2. _____ Select your fixed PLC from the Select Processor Type window.
3. _____ Verify the proper driver is selected.
4. _____ Click on OK to exit this window.

CONFIGURING THE LADDER WINDOW DISPLAY

RSLogix 500's ladder window can be customized to suit the programmer. The following portions of the ladder window can be customized:

- The type, font, style, and size can be selected by the user.
- Address display properties can be configured.
- Documentation can be displayed or turned off to allow more room to display ladder rungs.
- The formatting of the documentation can also be selected.
- I/O types can be displayed on ladder diagrams.
- Ladder instructions can be seen in three-dimensional format.
- Quick Key Mapping can be set up.

Either right-click on a blank spot on the ladder window to select Properties, or click on View and select Properties to display the View Properties window. The tabbed window opens as illustrated in Figure 10-28.

LAB TWO

This lab exercise will familiarize you with the View Properties tabs. The ladder properties will be set up and modified in the documentation lab exercise.

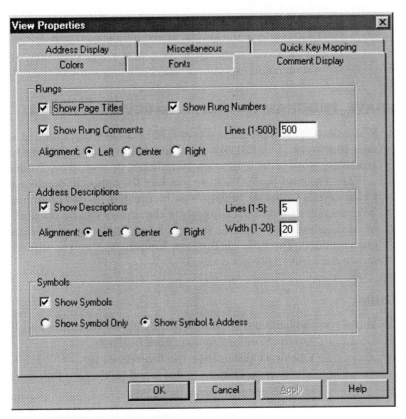

Figure 10-28   View Properties window.

1. _____ Click on View and select Properties to open the View Properties window.
2. _____ Click on each tab and familiarize yourself with the contents and format of each tab as their contents are described below.

Comment Display tab.
   Show page titles
   Show rung numbers
   Show rung comments
   Show address descriptions
   Show symbol only or show symbol and address.

Colors tab. The Colors tab is used to set up colors for the objects displayed on the ladder window. As an example, the background colors for addresses, page titles, symbols, comments, and power flow can be selected.

Fonts tab. The font size displayed on the ladder window can be selected in this tab.

Address Display. Addresses displayed on the ladder window can be formatted under this tab. Bit address format, binary bit display mode, I/O display, values for indirect addresses, and cross-reference display are set up in the Address Display window.

Miscellaneous tab. The Miscellaneous tab is where rung wrapping, page headers, display I/O types, and full drag properties options are selected.

Quick Key Mapping tab. There are many ways to program ladder logic instructions on a ladder rung. Quick Key Mapping is a feature that allows a programmer to assign a PLC programming instruction to each alpha key on the personal computer's keyboard. The assigning of an instruction to a keyboard key is called mapping. This tab is where the keyboard mapping is configured.

3. _____ When you begin developing ladder logic you may wish to return to the Properties screen and experiment with changing some of the options.

AUTO SAVE, PROGRAM BACKUP, AND REVISION NOTES

When developing a program, it is easy to become wrapped up in what you're doing and forget to save your work. The RSLogix AutoSave feature can be set up to automatically save your programming work every 5 minutes, 10 minutes, or even every 1 minute.

The Program Backup feature provides the option to create a backup copy each time the project is saved. Up to 99 backups can be created. If the programmer wishes to discard current program edits, the last backup copy of the project can be opened.

As a maintenance worker, it would be nice to have a place to put a note each time a PLC program is edited. The RSLogix Revision Notes feature can be turned on to display a Revision Notes screen each time a project is saved.

LAB THREE

This section will guide you through setting up the AutoSave and Revision Notes features.

1. _____ On the Windows toolbar, click on Tools.
2. _____ Click on Options from the drop-down list.
3. _____ The System Options dialog box should open. Figure 10-29 illustrates the top portion of the System Options dialog box.

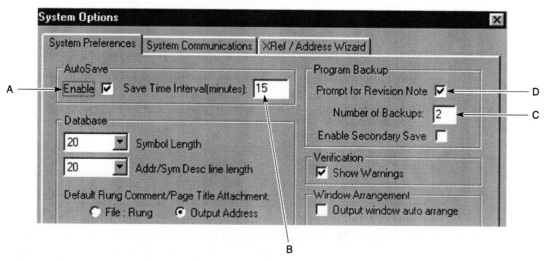

Figure 10-29 System Options dialog box.

4. _____ From the System Preferences tab, click on the Enable check box under AutoSave to enable the AutoSave feature. Refer to (A) in Figure 10-29.
5. _____ Enter the value in the text box (B) for the time interval between auto saves.
6. _____ Enter the number of auto backups that will be made in the text box (C) in Figure 10-29.
7. _____ Click on the check box (D) to enable a Prompt for Revision Note each time the project is saved.
8. _____ Click Apply.
9. _____ Click OK to leave the System Options dialog box.
10. _____ In order to turn on the AutoSave feature, the project must be saved. Save this project and name it Begin.
11. _____ Click on File on the Windows menu bar.

12. _____ Click on Save As . . . .
13. _____ Click on the "Save in" arrow and select the desired drive. Refer to (A) Figure 10-30.

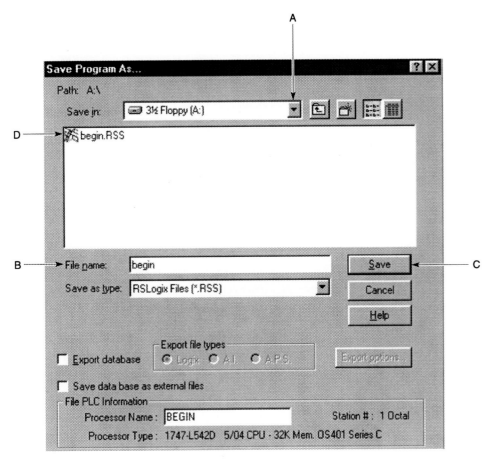

Figure 10-30 Save Program As . . . dialog box.

14. _____ Type in the file name Begin as illustrated in (B) Figure 10-30.
15. _____ Click on Save to save the project (C). Since this is the first time this project has been saved, a revision note in the dialog box will not display.
16. _____ Callout (D) in Figure 10-30 illustrates the Begin file as it will be displayed next time you either open a file or perform a save function.

Figure 10-31 illustrates a revision note for future saves of this project.

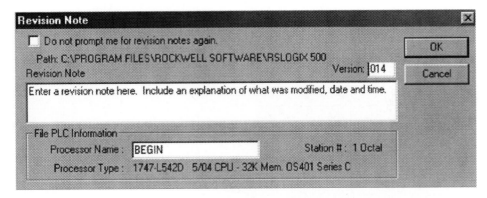

Figure 10-31 RSLogix 500 Revision Note window with a sample revision note.

As you work on developing ladder logic, the Program Backup feature will make a backup of the project every time the project is saved. Figure 10-32 illustrates the Open File/Import SLC 500 Program dialog box. Notice the Begin project and three backup files listed directly below. BEGIN_BAK000 is the first backup file. Figure 10-32 shows backup files 0 through 2. When another save is made, backup 000 will be discarded and backup 003 will appear.

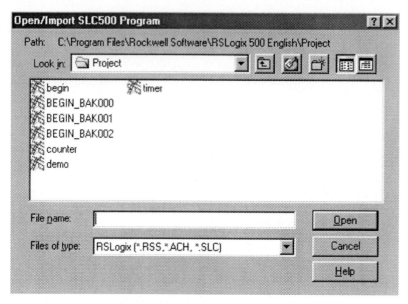

Figure 10-32   Open/Import SLC 500 Program dialog box showing auto backup files.

11

Introduction to ControlNet and DeviceNet

LAB 11A: CONTROLNET NETWORK

1. ControlNet is an open control network used for real-time data transfer of _____ and _____ data between processors or I/O on the same link.

2. The _____ is the bus or backbone of the network.

3. The trunk cable may be comprised of one or multiple sections called _____.

4. Up to _____ nodes may be on a segment before a repeater must be used.

5. Explain the difference between time-critical and non-time-critical data transfers.

6. When setting up an application, how does the time-critical data differ from the non-time-critical data?

7. The ControlNet data transfer rate is _____ bits per second.

8. Up to _____ nodes can be connected on a ControlNet network.

9. ControlNet is an open network managed by _____.

10. Depending on the environment in which your cable is installed, the trunk line is either quad-shielded _____, _____, or special-use cable.

11. Because ControlNet is an open network, hardware from _____ can be purchased and connected as a node to the network.

12. When designing a redundant ControlNet network, it is important that cables be routed _____ so that if one cable is damaged, the other cable remains intact and can continue communications.

13. The main cable is called the _____.

14. Taps are required for connecting the _____ to each individual node. Maximum drop line length is _____ inches.

15. If using fiber and the appropriate repeaters, a ControlNet network can be up to _____ long.

16. The 1747-SCNR scanner is used to provide scheduled and unscheduled messaging between devices on ControlNet and the _____.

17. The ControlNet _____, referred to simply as the NAP, is an RJ45 connection point for a personal computer interface with the network.

18. The 1756 CNB or 1756 CNBR is the ControlLogix PLC interface module to the ControlNet network. CNB stands for _____.

19. The CNB has only _____ network connection, whereas the CNBR contains _____ connectors for network _____.

20. A _____ card is a PCMCIA credit card-sized interface card for inserting into a notebook computer's PCMCIA slot. It is used to interface with the ControlNet network.

21. The maximum length of a segment is defined as:

22. Refer to your textbook and identify the labelled components of the ControlNet network illustrated in Figure 11-1.
    A.

    B.

    C.

    D.

    E.

    F.

    G.

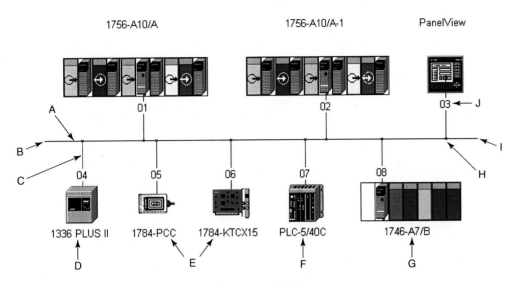

Figure 11-1 ControlNet network as viewed with RSNetworx.

H.

I.

J.

23. On the following page, identify the module in and features of Figure 11-2.

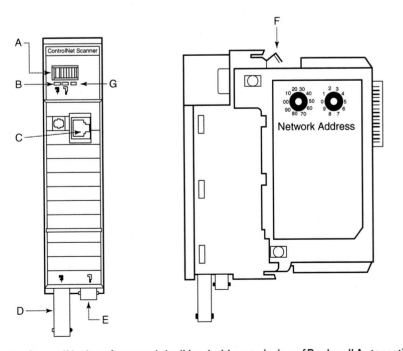

Figure 11-2 ControlNet interface module. (Used with permission of Rockwell Automation, Inc.)

A.

B.

C.

D.

E.

F.

G.

24. The following questions refer to Figure 11-3.

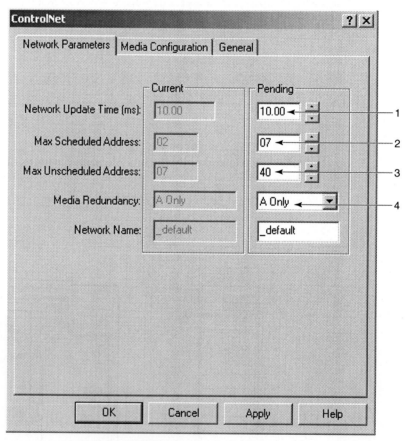

Figure 11-3   ControlNet setup screen.

A. What software is this screen from?

B. Define the four identified parts shown in Figure 11-3.
    1.

2.

3.

4.

25. What is the formula for calculating the maximum length of a ControlNet segment?

26. You need to install a ControlNet network that is 2,500 meters between the first and last nodes. There are a total of 34 devices to be placed on the network.
    A. Can this network be installed as a single segment?

    B. Even though there are many answers to this question, how many segments would you suggest? Make a rough sketch of the network layout.

    C. Show your calculations below.

    D. How many repeaters would you need?

    E. How many termination resistors are required?

27. What is the maximum length of a segment that requires 18 taps?

28. Illustrate and identify the pieces of the NUT bandwidth shown in Figure 11-20 of the textbook.

29. Configure a ControlNet network using RG6 coax cable with the following node information:
    - Twelve nodes are needed for transmitting scheduled data only.
    - Five additional nodes are needed for transmitting scheduled as well as unscheduled data.
    - Three additional nodes are needed for future use. They will transmit scheduled (and possibly unscheduled) data.
    - Five nodes are required for current unscheduled ControlNet service.
    - Three additional nodes are needed for NAP connectivity.
       A. Can this configuration be accomplished with one segment?

       B. What is the maximum length of the coax segment?

       C. How many taps are required?

D. How many terminating resistors are required?

E. What software are you going to use to configure these parameters?

F. What value are you going to program into SMAX?

G. What value are you going to program into UMAX?

H. What node addresses are available for the NAP?

I. You are using a 1756 CNB module. What will you program into the Media Redundancy parameter?

J. After programming these parameters, what must be done to update the actual hardware?

K. How many nodes are going to fall within slot time and why?

L. After the network is installed and running, a maintenance worker wants to connect to the ControlNet network at the NAP on a CNB with his personal computer. He has a 1784-PCC ControlNet interface card in his notebook computer. Configuring the RSLinx driver and accepting the default MAC address makes his personal computer node 99 on the ControlNet network. Is communication between the personal computer and the ControlNet network possible? Why or why not? Explain your answer.

M. If the answer to the above question is "no," and the maintenance worker came to you for help, what would you do?

LAB 11B: DEVICENET

1. DeviceNet is an open network intended to link low-level devices to higher-level devices such as PLCs. List a few devices one might find on DeviceNet.

2. DeviceNet, similar to ControlNet, is managed by an independent group called the _____, or ODVA.

3. List the components that make up a DeviceNet network.

4. Each end of the trunk line is required to have one _____-ohm _____-watt terminating resistor.

5. The cable has four wires: two for _____ and two for _____.

6. Depending on the total amount of data transferred between the nodes, a DeviceNet network can support up to _____ decimal nodes.

7. When designing a new network or adding additional devices to the network, _____ calculations must be completed in order to ensure proper power supply selection and network operation.

8. Node commissioning consists of taking a new device and setting up its _____ and _____, and programming its _____.

9. List the two different ways node commissioning can be completed.

10. One very important consideration for on-network live commissioning is that _____ addresses are not allowed.

11. Many devices come from the factory as node _____; however, some newer hardware- and software-configurable devices are shipped with an address of _____.

12. What equipment is required for setting up a node commissioning station?

13. Electronic data sheets, or _____, contain information regarding the personality of the device.

14. The correct EDS file must reside within each _____ before it can become a working part of the DeviceNet network.

15. The EDS file must match the _____ level of the network device.

16. If the EDS file is not current, you may have to go on the Internet to either the manufacturer's or the _____ web site and download the correct file to your personal computer.

17. Once the file is on your personal computer, use the _____ wizard to update or register the network device.

18. Fill in the table shown in Figure 11-4.

DeviceNet Cumulative Drop Line Length	
Baud Rate In Bits Per Second	Cumulative Length Allowed

Figure 11-4 DeviceNet Cumulative Drop Line Length.

19. Fill in the table shown in Figure 11-5.

Maximum Trunk Line Cable Length			
Baud Rate In Bits Per Second	Thick Round	Thin Round	Flat Cable

Figure 11-5 DeviceNet Maximum Trunk Line Cable Length.

20. Perform the following list of network cable calculations for the DeviceNet network shown in Figure 11-6. Assume that the cable is round and thick. The numbered arrows refer to cable lengths. All lengths are in feet.

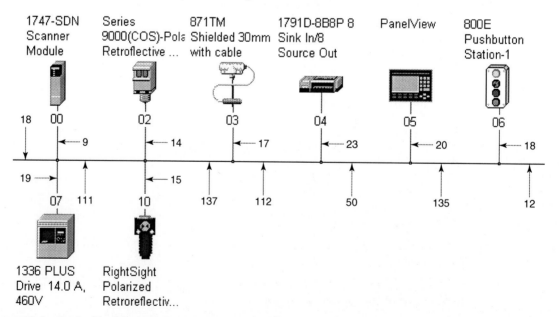

Figure 11-6 DeviceNet lab trunk line calculation example.

What is the cumulative trunk line length?

What is the cumulative drop line length?

Review your calculations. Do you see any problems with your network configuration?

What is the maximum baud rate possible for this network?

If you needed to have a 500K baud rate, what would you have to do?

ADDITIONAL CONSIDERATIONS REGARDING TRUNK LINE CALCULATIONS

When performing trunk line calculations, there is a slight complication of which we need to be aware. It is important to calculate the entire length of the trunk line. In order to accomplish this, we have to look at our network a bit differently.

Notice that in Figure 11-7, the distance between the terminating resistor on the left end of the network and node 0 is 3 feet; the distance between node 0 and 1 is 4 feet; and the length of the drop line to node 7 and the terminating resistor is 15 feet. Since the drop line is longer, this length (15 feet) should be used when calculating the total trunk line length, not the (7-foot) trunk line length between node 7 and the terminating resistor.

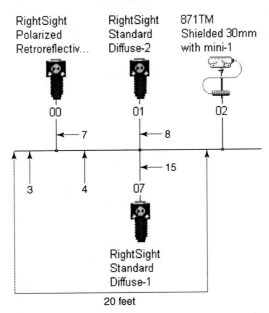

Figure 11-7   Maximum trunk line cable length calculation including drop line.

So the rule for performing trunk line calculations is: Move 20 feet away from a terminating resistor in order to see if there is a drop line that is longer than the actual trunk line. If there is, add the drop line length, not the trunk line length, to the total calculation of the trunk line length.

1. Now go back and look at your calculations for Figure 11-6. Do you now have a different answer for the total calculation of the trunk line length?

Will this new answer change the baud rate?

Can you see how it might change the baud rate?

2. Calculate the trunk line cable length for the network in Figure 11-8. Assume the network uses a thick, round cable.

What is the maximum baud rate for this network?

What is the cumulative drop line length for this network?

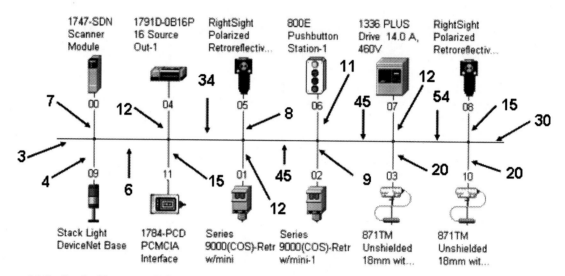

Figure 11-8   DeviceNet network for trunk line calculation.

DEVICENET NETWORK POWER SUPPLY CALCULATIONS

When designing or modifying a DeviceNet network, the power supply size needs to be calculated. The power supply can be end-mounted or mounted in the center of the network. The best method is to center-mount the power supply; however, end-mounting is acceptable. In some cases, a second power supply may be required. As the network is expanded and additional power is required, a second power supply may need to be added in order to satisfy the power requirements for the network nodes.

Example: A DeviceNet network in Figure 11-9 has the following characteristics:

Thick, round cable
A trunk line length of 722 feet
One end-mounted power supply

There are 8 nodes on the network:

1747-SDN DeviceNet interface card at node 0
1784-PCD Computer interface card at node 11
1791D-OB16P Compact Block I/O as node 4
1336 PLUS Drive, node 7
800E Push Button Station at node 6
855T Stack Light at node 9
2- RightSight Photo Sensors at node 5 and 8

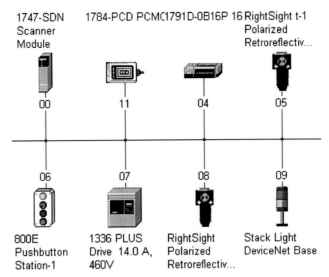

Figure 11-9   Sample DeviceNet network for calculation.

Problem Solution for Figure 11-9:

The round, thick trunk line cable is 722 feet and has a single end-mounted power supply. Adding up the total current required using the table in Figure 11-10, we have a total of 890 mA. Looking at the Maximum Current Table in Figure 11-11, at 722 feet we can draw up to 1.39 amps of current to the network.

Each node's current requirement is listed in Figure 11-10:

DeviceNet Node Power Requirement:	
DeviceNet Node Device:	DeviceNet Current:
RightSight Photo Sensor	60 mA
Series 9000 Photo Sensor	70mA
1336 PLUS Drive DeviceNet Interface	40mA
1791D-0B16P	200mA
871TM Inductive Proximity Sensor	60mA
800 E Push Button Station	50mA
1747-SDN Scanner	90mA
1784-PCD Personal Computer Interface	90mA
855T Stack Light	300mA

Figure 11-10   DeviceNet node power requirements.

Figure 11-11 lists the network cable length and associated maximum allowable currents.

LAB EXERCISE A

Calculate the required power supply for the DeviceNet network shown in Figure 11-12. Assume the network uses the following:

A thick, round cable

A single end-mounted power supply

One End-Mounted Power Supply Using Round, Thick Cable	
Network Length:	Maximum Amps:
262 feet/80 meters	3.59
328 feet/100 meters	2.93
394 feet/120 meters	2.47
459 feet/140 meters	2.14
525 feet/160 meters	1.89
591 feet/180 meters	1.69
656 feet/200 meters	1.53
722 feet/220 meters	1.39
787 feet/240 meters	1.28
853 feet/260 meters	1.19
919 feet/280 meters	1.10

Figure 11-11   Partial list of DeviceNet currents derived from Rockwell Automation tables.

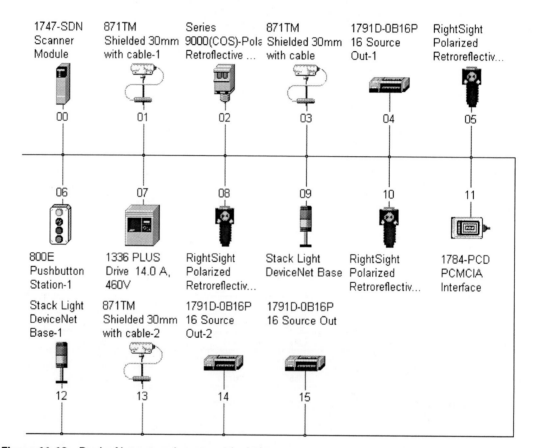

Figure 11-12   DeviceNet network power calculations.

Determine the trunk line cable length in Figure 11-12, Figure 11-13 (use column for Lab A), and Figure 11-14.

Trunk line cable length: _____ feet

Maximum allowable current: _____ amps

Current required by network devices: _____ amps

Network Cable Lengths:		
Between nodes:	Distance in Feet for Lab A	Distance in Feet for Lab B
Terminating resistor and node 0	10	12
Node 0 and 1	78	90
Node 1 and 2	56	50
Node 2 and 3	88	88
Node 3 and 4	62	64
Node 4 and 5	9	35
Node 5 and 15	8	18
Node 15 and 14	55	43
Node 14 and 13	72	75
Node 13 and 12	8	30
Node 12 and terminating resistor	6	12

Figure 11-13   Network cable lengths.

Drop Line Lengths	
From Trunk Line to Node:	Drop Line Length in Feet;
0, 2, 5	20
6, 12	15
14	12
3, 8, 15	10
4, 9, 13	8
1, 10, 11	6

Figure 11-14   Network drop line lengths.

LAB EXERCISE B

Calculate the required power supply for the DeviceNet network in Figure 11-12. Assume the network uses the following:

A thick, round cable

A single end-mounted power supply

Determine the trunk line cable length in Figure 11-12, Figure 11-13 (use column for Lab B), and Figure 11-14.

Trunk line cable length: _____ feet

Maximum allowable current: _____ amps

Current required by network devices: _____ amps

RSNETWORX FOR DEVICENET LAB EXERCISES

The next section will pose questions about the information found in RSNetWorx for DeviceNet software screens.

The Following Questions Refer to Figure 11-15

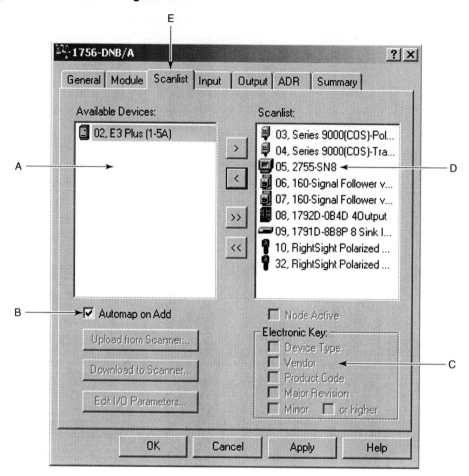

Figure 11-15 1756 DNB RSNetWorx for DeviceNet.

1. Identify A.

2. Identify B.

3. Explain the options available for C.

4. Why do the listed items appear in the window (D)?

5. E refers to the Scanlist tab. Explain the importance of the Scanlist.

The Following Questions Refer to Figure 11-16

6. A is pointing to the part number 1756-DNB. What is this window heading telling you?

7. H identifies the Input tab. Describe what it does.

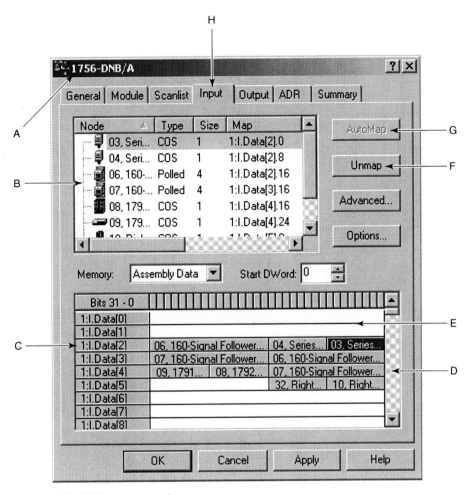

Figure 11-16   1756-DNB input mapping.

8. Describe the contents of the window identified by B.

9. What is C pointing to?

10. D is pointing to _____ .

11. Why is there no data in the area identified by E?

12. What would happen if you clicked on F?

13. Assume the button identified by G is active. What would happen if you clicked on it?

The Following Questions Refer to Figure 11-17

14. G is pointing to 1747-SDN. What does this signify?

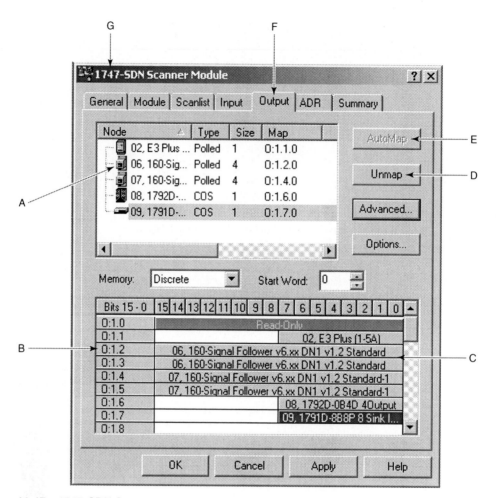

Figure 11-17   1747-SDN Scanner output mapping.

15. Describe the contents of the window identified by A.

16. What is B pointing to?

17. C identifies a Bulletin 160 variable frequency drive.

   The drive is what node on the DeviceNet network?

   How many data words are mapped to this drive?

   What are the output addresses mapped to this drive?

   When the 1747-SDN is read by the I/O configuration in the RSLogix 500 software, how many output words are assigned to that particular chassis slot and the module?

18. The mapping at O:1.7 is highlighted. What will happen if you click on the button pointed to by D?

19. E is pointing to AutoMap. What is the disadvantage of AutoMapping?

20. What is the difference between the Input tab and the Output tab (F)?

The Following Questions Refer to Figure 11-18

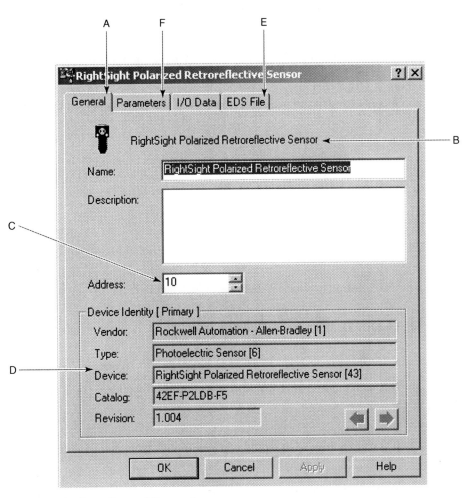

Figure 11-18   RightSight General Properties page.

21. Identify A.

22. What does B tell you?

23. Identify C.

24. Identify D.

25. E is pointing to the EDS File tab. What does EDS stand for?

    Why is the EDS file important?

    If you do not have a current copy of the EDS file, where can you get one?

    After you have downloaded an updated copy of an EDS file, how do you download it into the device?

    In what radix will the downloaded EDS file be stored on your personal computer?

26. F is pointing to the parameters tab. Why would you click on this tab?

The Following Questions Refer to Figure 11-19

27. Identify the tab referred to by A.

28. B is pointing to the number 10. What does this number signify?

29. C is pointing to a lock. What does a lock signify?

30. Identify D. What is this field used for?

31. If E were active, what would you accomplish by clicking on it?

32. If F were active, what would you accomplish by clicking on it?

33. What does G identify?

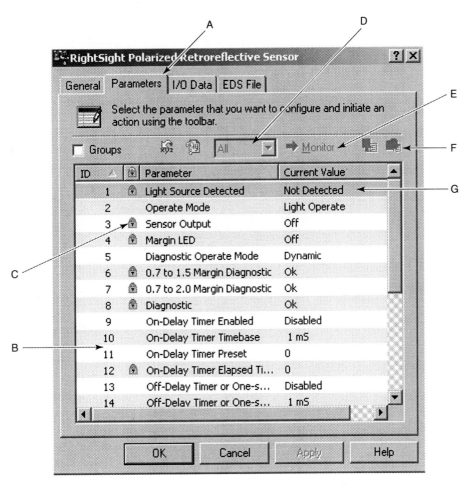

Figure 11-19   RightSight Photo Sensor Parameters page.

EDS File Determination Lab

In this lab, we are going to use the General Properties screen device identity information to determine the EDS file hexadecimal value. You need to use the EDS Wizard to update the EDS files in each device we are using in our PLC program. From the General Properties screen for each node used in our SLC 500 project, determine the EDS file number.

34. Determine the EDS file number for the 1792D-OB4D Armor Block illustrated in Figure 11-20.

35. Determine the EDS file number for the 2755-SN8 Adapta Scan bar code reader from the information provided in Figure 11-21.

36. Determine the EDS file for the 1747-SDN scanner shown in Figure 11-22.

37. Determine the EDS file for the 1791D Compact Block I/O module shown in Figure 11-23.

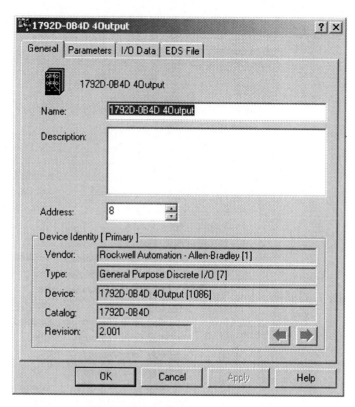

Figure 11-20   Armor Block 1792D-OB4D General Properties page.

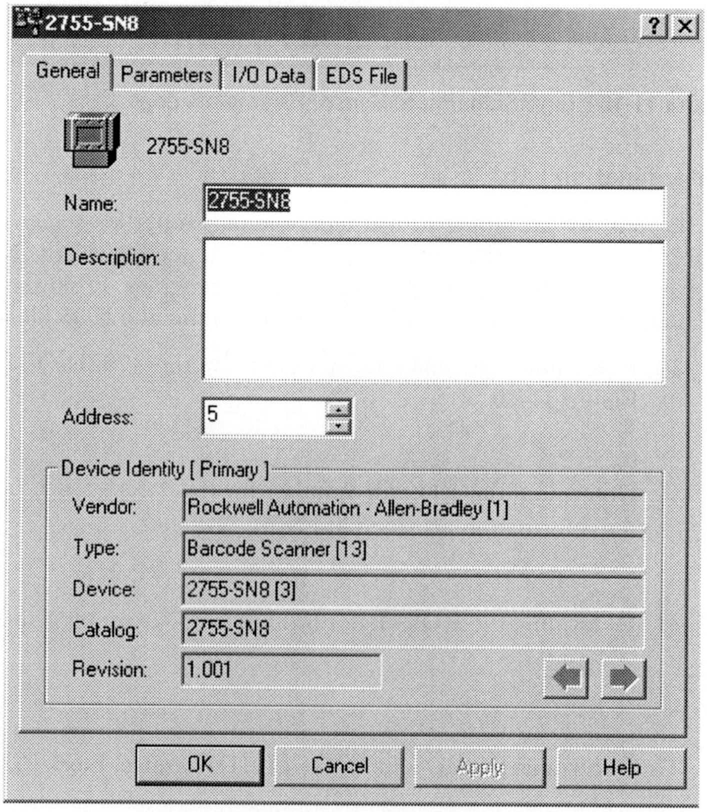

Figure 11-21   2755-SN8 Adapta Scan barcode scanner General Properties page.

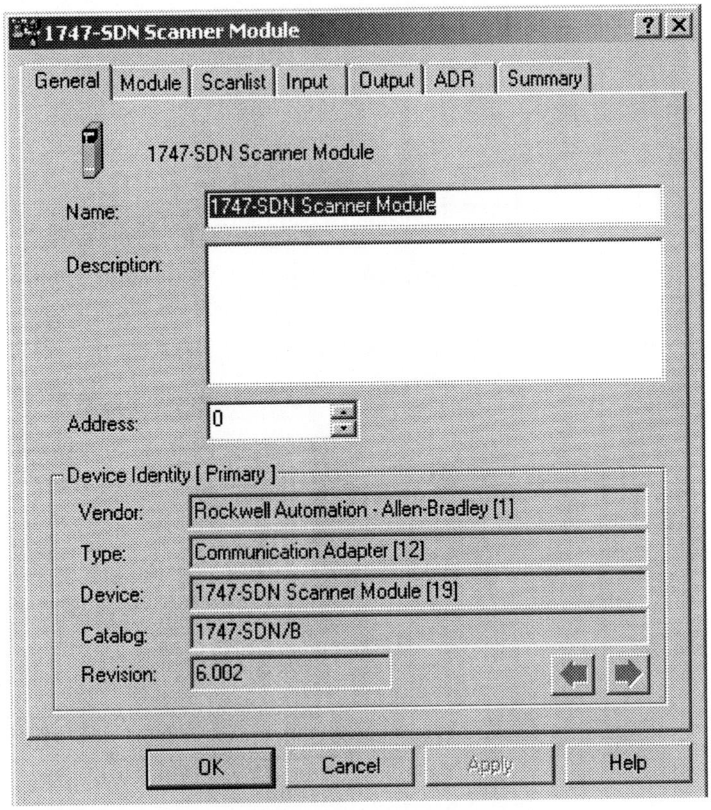

Figure 11-22   1747-SDN scanner General Properties screen.

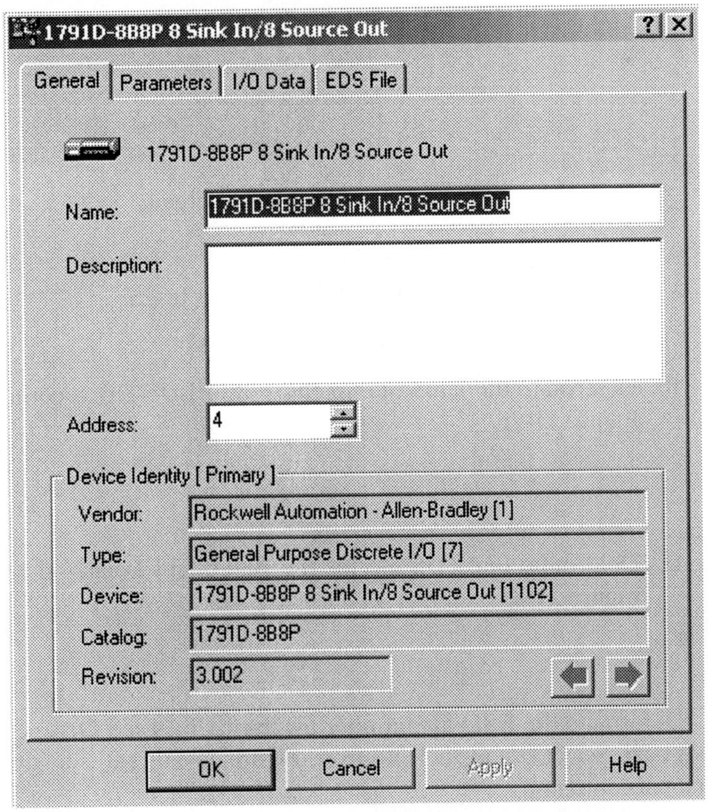

Figure 11-23   1791D Compact Block I/O General Properties screen.

38. Determine the EDS file for the 800E Pushbutton station shown in Figure 11-24.

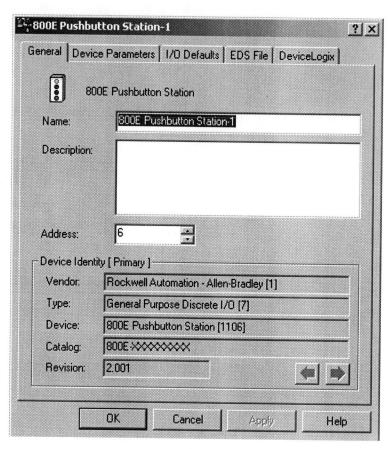

Figure 11-24 800E General Properties screen.

39. Select the correct EDS file from Figure 11-25.

40. Determine the EDS file for the 1336 PLUS Drive communications interface shown in Figure 11-26.

41. Determine the EDS file for the RightSight Polarized Retroreflective Sensor shown in Figure 11-27.

Determine the ESD file number for the 1336 PLUS Drive communication interface in Figure 11-28. Select the proper file from Figure 11-29 that is a copy of selected EDS files from a personal computer.

Figure 11-25 Partial list of EDS files from personal computer.

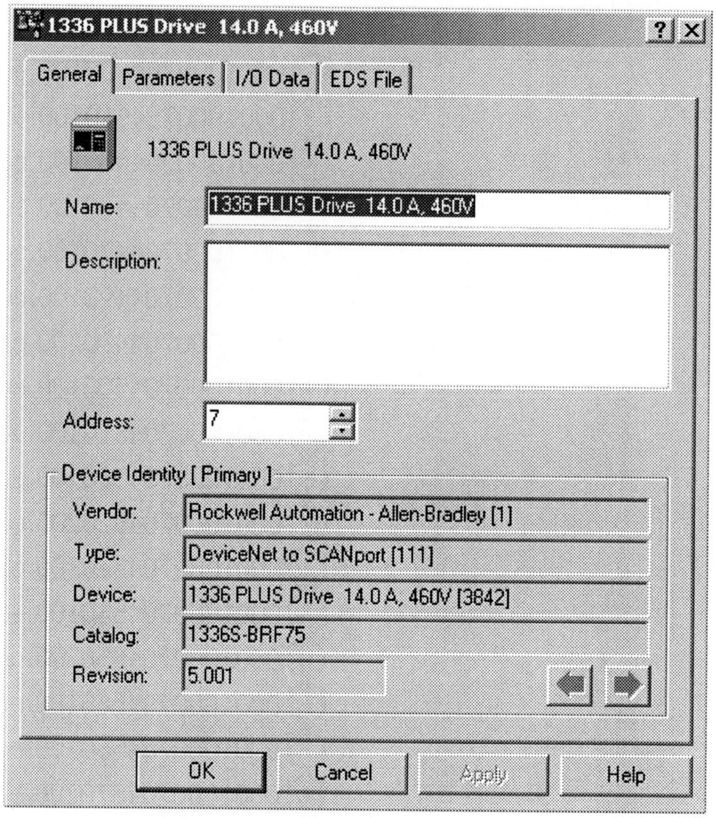

Figure 11-26 1336 PLUS Drive General Properties screen.

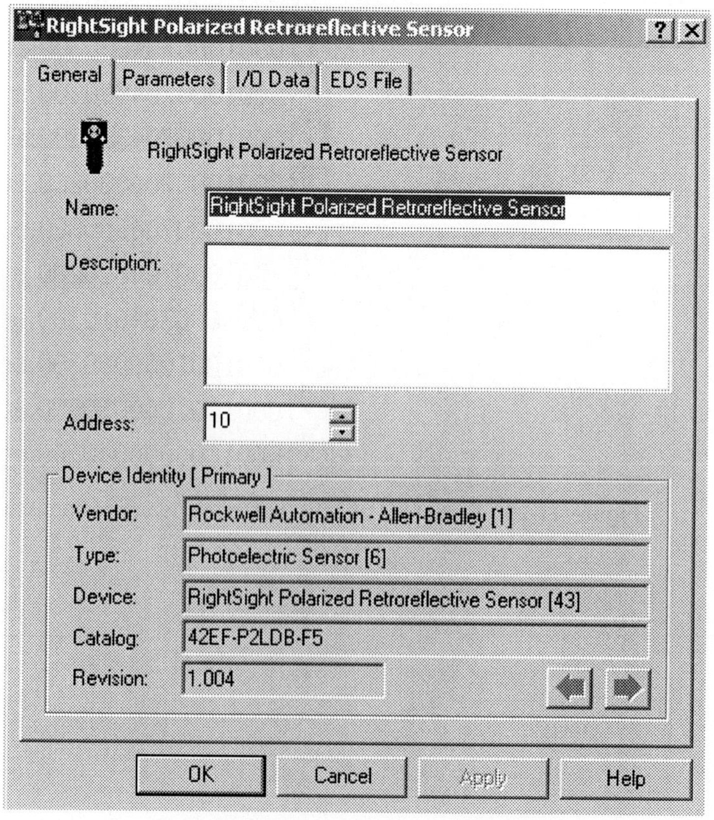

Figure 11-27 RightSight Retroreflective Sensor General Properties page.

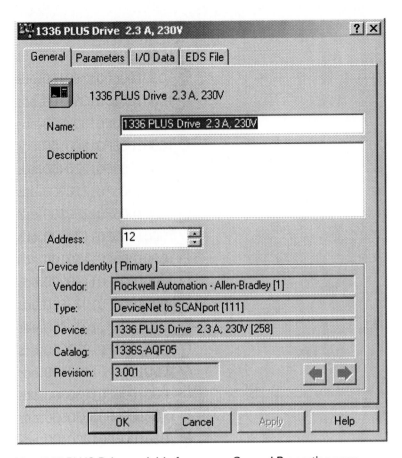

Figure 11-28   1336 PLUS Drive variable frequency General Properties page.

Figure 11-29   Partial list of EDS files from personal computer.

LAB EXERCISE

12

Working with Files and Addresses

PREREQUISITE

Before completing the lab exercise, read Chapter 12, Data Organization, in the textbook.

OBJECTIVES

Upon completion of this laboratory exercise, you should be able to:

- determine input and output addresses
- identify data and its associated address in different data files
- enter data into a data file
- expand a data file
- create a user-defined data file
- modify data in a data file
- identify a file within a data file

INTRODUCTION

This portion of the file and address lab will provide experience identifying data file addresses and contents from RSLogix 500 data file screen prints. We will work with identifying data in the output status, input status, bit, integer, and floating-point files. We will look at timer and counter data files in their respective sections. The data files are displayed on the project tree from Figure 12-1.

LAB ONE

Output Status File

The output status file has one bit to represent each output screw terminal on the SLC 500 PLC. The screw terminal addresses are dictated by the slot in which the output module resides in the modular PLC chassis. Figure 12-2 illustrates the RSLogix 500 output status file. Using information in the file, answer the questions on the next page.

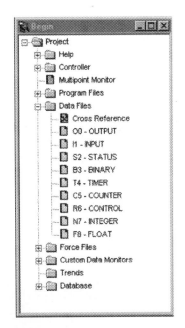

Figure 12-1 The Data File folder expanded in the RSLogix 500 project tree for the project named Test.rss.

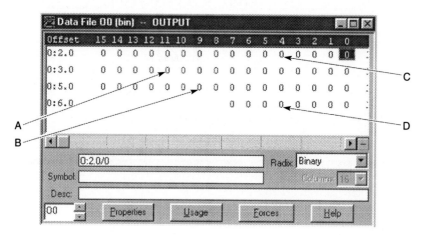

Figure 12-2   RSLogix 500 Data File 0, Output status file.

1. _____ Which chassis slots does the status table identify as output modules? _____

_____

2. _____ Why does the data word associated with O:6.0 have less bits than other words in the

file? _____

3. _____ List addresses below as identified in Figure 12-2.

    A. _____

    B. _____

    C. _____

    D. _____

Input Status File

The input status file has one bit to represent each input screw terminal on the SLC 500 PLC. The screw terminal addresses are dictated by the slot in which the input module resides in the modular PLC chassis. Figure 12-3 illustrates the RSLogix 500 input status file. Using information in the file, answer the questions below.

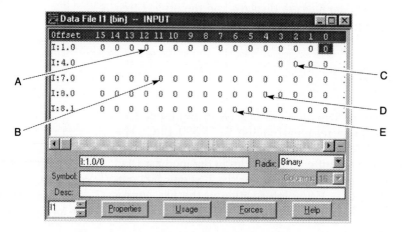

Figure 12-3   RSLogix 500 Data File 1, Input status file.

1. _____ Which chassis slots do the data file or status table identify as input modules?

_____

2. _____ Why does the data word associated with I:4.0 have less bits than other words in the file? _____

3. _____ What do I:8.0 and I:8.1 signify? _____

_____

4. _____ Slot 8 uses how many words to store its input data? _____
5. _____ List addresses below as identified in Figure 12-3.

A. _____

B. _____

C. _____

D. _____

E. _____

Binary File

The binary file is also called the bit file. This file is used to store binary 1s and 0s signly or as a group. A typical application using the bit file would be to store an output bit that would not control real-world outputs, but rather other ladder rungs in your program. These storage bits are sometimes called internal bits, internal relays, or internal coils. Using a binary storage bit eliminates the need to use a real output point to store internal control bits. A second use for the binary file would be interfacing an operator interface device such as an Allen-Bradley Panel View operator interface terminal to an SLC 5/04 processor using the Data Highway Plus network. Since the Panel View is a touch-screen operator interface device, it is not possible to attach input or output wires to the screen objects and connect them to an input or output module. When we develop our Panel View screens, binary bit addresses (called tags) will be assigned to each touch-screen object. These addresses will be used to correlate screen objects to ladder logic.

Figure 12-4 illustrates the RSLogix 500 Bit file B3 file. Using information in the file, answer the questions below.

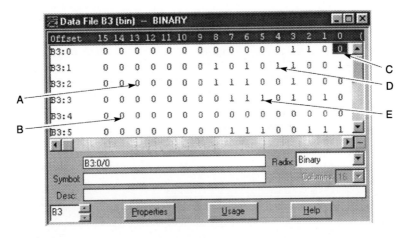

Figure 12-4   RSLogix 500 Data File 3, Bit or Binary file.

1. _____ List addresses below as identified in Figure 12-4.

A. _____

B. _____

C. _____

D. _____

E. _____

Integer File N7

The integer file contains whole numbers. Data in an integer file could be stored as the answer to a math operation, as production data from counters, and as operator entered data, as well as for recipe storage.

Figure 12-5 illustrates the RSLogix 500 Integer File N7. Using information in the file, answer the questions below.

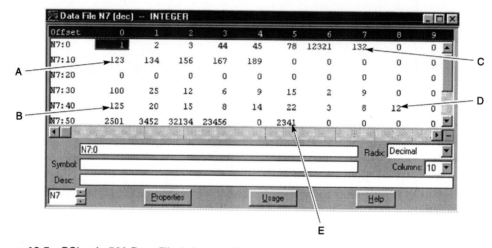

Figure 12-5   RSLogix 500 Data File 7, Integer file.

1. _____ What kind of data is stored in the integer file? _____

_____

2. _____ What is the range of data allowed in an integer file? _____

3. _____ Can fractions such as .25 be stored in an integer file? _____

4. _____ When we say we have a file within a file what do we mean? _____

_____

5. _____ A chocolate chip cookie recipe is stored in integer file N7 starting at element N7:30. This recipe has a length of 8 elements. What is the address of the last element? _____

6. _____ List addresses and data stored there below as identified in Figure 12-5.

A. _____   _____

B. _____   _____

C. _____   _____

D. _____   _____

E. _____   _____

Floating-Point File

The floating-point file allows numbers greater than 32,767 and fractional values such as .5 or .25 to be used with the SLC 500 family of modular processors. Floating-point files are only available in modular processors and the MicroLogix 1200 and 1500 with the proper operating system.

Figure 12-6 illustrates the RSLogix 500 Floating-Point File F8. Referring to information in the file, answer the questions.

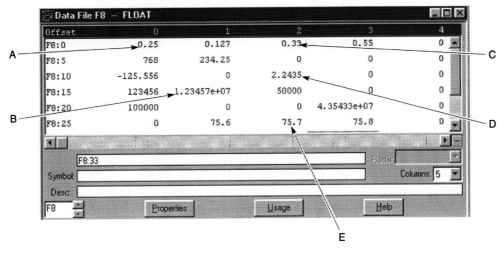

Figure 12-6   RSLogix 500 Data File 8, Floating-Point Data file.

1. _____ What kind of data is stored in the floating-point file? _____

_____

2. _____ The floating-point file is processor operating system dependant. What does this

mean? _____

_____

_____

3. _____ In order to use the floating-point file you must have a minimum of a _____

modular processor with _____ .

4. _____ The number 12,345,700 is stored at what address? _____

5. _____ The value .25 is stored at what address? _____

6. _____ Expand the number stored in F8:23 into a decimal number. _____

7. _____ What does the e+07 mean in a floating-point value? _____

_____

_____

8. _____ Describe the difference between an integer file and the floating-point file. _____

_____

_____

9. _____ List addresses and data contained below as identified in Figure 12-6.

A. _____   _____

B. _____   _____

C. _____   _____

D. _____   _____

E. _____   _____

LAB TWO

This lab will provide you experience working with the software and correlating the field input devices and their addresses that were wired up to the input status table bit positions in lab exercise 8.

1. _____ Your instructor should have your RSLogix 500 software running the Begin project on the lab computers.
2. _____ The input field devices wired in lab 8 should be hooked up and ready for use.
3. _____ In the table in Figure 12-7, list the input field devices and the addresses you determined they were connected to in lab exercises 8 and 9.

FIELD INPUT DEVICES INTERFACED TO INPUT MODULES		
Input Device Type	Input Address	Find Address in Data Table?

Figure 12-7   Input field devices and their input addresses.

4. _____ From the project tree in the RSLogix 500 software, expand the data files folder if it is not already expanded.
5. _____ Double-click on the input data file to display it.
6. _____ Verify that your PLC processor is in remote run mode (REM Run).
7. _____ View the input data table as you energize and de-energize each input device. Verify that the expected input address correlates with the input status file bit as it transitions from a 0 to a 1.

LAB THREE

This lab will also take you to the RSLogix 500 software. This lab will look into RSLogix 500 data files and view data, expand the files to accept additional data, and enter and modify current data.

1. _____ Open your Begin project if it is not currently open.
2. _____ Go to the project tree and verify that the Data File folder has been expanded.
3. _____ Double-click to open the Binary, or Bit file.
4. _____ Click on Properties.
5. _____ Expand this file to contain 50 elements.
6. _____ Click on OK.
7. _____ Close the Binary file.
8. _____ Open the Integer file.
9. _____ Expand this file to 50 elements.
10. _____ Select element N7:22.

11. _____ Enter the value 1234.

12. _____ Enter the following integers in the following addresses:

N7:2	54
N7:42	157
N7:15	2500
N7:33	30250

13. _____ Enter the following recipe data into the file starting at N7:10: 100, 25, 14, 7, 34, 18, 1, 3.

14. _____ What is the length of this recipe file? _____

15. _____ What is the last address of this recipe file? _____

16. _____ Enter the value 40000 into address N7:5.

17. _____ Explain what you observe when you enter this value. _____

_____

_____

Skip to 24 if your PLC does not support floating-point files.

18. _____ Open the Floating-Point file.
19. _____ Expand the file to 25 elements.
20. _____ Enter the value .25 in F8:12.
21. _____ Enter the value 1.2375 in F8:3.
22. _____ Enter the value 40000 into F8:9.
23. _____ Enter the value 1234567.
24. _____ If you have a MicroLogix 1000, skip to the review questions.

Creating a User-Created Data File

Data files 0 through 8 are created as part of the opening of a new project. Modular processor data files may be expanded up to 256 data files, files 0 through 255, assuming adequate processor memory.

> NOTE: If you are using a fixed PLC, you will not be able to expand the number of data files. If you are using a fixed PLC or a 5/01, 5/02, or 5/03 modular processor with an operating system less than OS 302, floating-point files will not be available. The MicroLogix 1200 and 1500 controller will support floating-point files if you are using RSLogix 500 version 5.0 software, and if the MicroLogix is a Series C, firmware level 4.0.

1. _____ Right-click on Data Files (A) and New (B) as illustrated in Figure 12-8.
2. _____ Click on the down arrow next to File type as illustrated at callout (A) in Figure 12-9.
3. _____ Select Integer as the File type you intend to create.
4. _____ Type in a name for your new integer file; refer to (B) in the figure.
5. _____ Add a description of the file at (C).
6. _____ The file to be created will be file 9. This can be changed by typing a different number at (D).
7. _____ To create 25 integer elements, type 25 at (E).
8. _____ Click OK (F) when completed.
9. _____ Look in the Data File folder to see the new integer file, Integer file 9.
10. _____ Double-click on the new integer file to open it.
11. _____ Enter data into the following addresses:

N9:22	1754
N9:2	29500
N9:15	1436
N9:16	2154

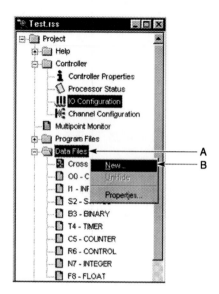

Figure 12-8 Opening a new data file from project tree.

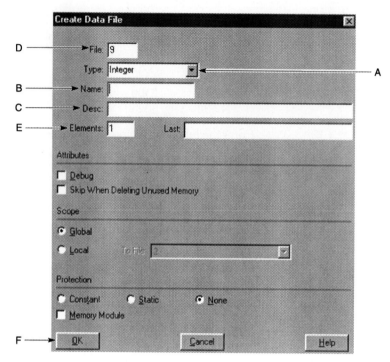

Figure 12-9 Creating a new data file.

12. _____ Click on Properties.
13. _____ Change the number of file elements to 20.
14. _____ Click OK when completed.
15. _____ What happened to the data in N9:22? _____
16. _____ Close this file.
17. _____ Right-click on Integer File N9.
18. _____ Click on Rename in the menu that opens.
19. _____ Rename the integer file.
20. _____ Press Enter when completed.
21. _____ Open a new Binary File 10.
22. _____ Name the file Panel View.
23. _____ Enter the file description as: operator interface inputs and outputs.
24. _____ Create 100 elements.
25. _____ Click OK to close the Create Data File window.
26. _____ Open a new Integer File 11.
27. _____ Name the file Line 1.
28. _____ Enter a file description as: production data line 1.
29. _____ Create 50 elements.
30. _____ Click OK to close the Create Data File window.
31. _____ Open a new Integer File 12.
32. _____ Name the file Line 2.
33. _____ Enter the file description as: production data line 2.
34. _____ Create 50 elements.
35. _____ Click OK to close the Create Data File window.
36. _____ Right-click again on Integer File N9.

37. _____ Delete Integer File N9.
38. _____ If your processor supports floating-point files, create a new floating-point file F13.
39. _____ Name the file Data.
40. _____ Create 20 elements.
41. _____ Add the following data to the listed addresses.

F13:0	123.25
F13:1	.333
F13:2	1.532
F13:4	.25
F13:5	47.5
F13:6	47.6
F13:7	47.7
F13:8	52500
F13:9	123,456
F13:10	72,394
F13:11	−2.7453
F13:12	.125

42. _____ When completed with the exercise, save it to your floppy disk.
43. _____ Close the Begin project.

REVIEW QUESTIONS

1. What is a word? _____
2. How many bits make up an SLC 500 memory word? _____
3. Define bit. _____
4. Define file. _____
5. What is a processor file? _____
6. What is a program file? _____
7. What is the purpose of a data file? _____
8. In which other file is the data file located? _____
9. How many processor files may reside in the CPU at any one time? _____
10. How many program files can be contained in a processor file? _____
11. How many data files can be contained in a processor file? _____
12. Identify what is contained in program files:

    File 0: _____

    File 1: _____

    File 2: _____

    File 3: _____

13. Identify what is contained in each data file in Figure 12-10.
14. Are all SLC address formats the same? _____ Why, or why not? _____

_____

File number	Identifier	File type
0		
1		
2		
3	B	
4		
5		Counter
6		
7		
8		

Figure 12-10   RSLogix 500 data files.

15. What are the three main items that make up an I/O address? _____

_____

_____

16. What data is contained in the input data files? _____

How is individual data represented? _____

Where does this data come from? _____

What does this data represent? _____

17. What are the three things that make up an input data file? _____

_____

One of these is used to represent input signals. How does it do that? _____

_____

_____

18. What are the three things that make up an output data file? _____

_____

One of these is used to represent output signals. How does it do that? _____

_____

_____

LAB EXERCISE

13

Lab A: Introduction to SLC Programming

OBJECTIVES

Upon completion of this laboratory exercise, you should be able to:

- understand the operation and programming of the examine-if-open and examine-if-closed instructions
- understand the operation and programming of the output energize instruction
- understand the operation and programming of the output latch and unlatch instructions
- perform minor program editing
- enter and save a revision note
- download, go on-line, and test the program

INTRODUCTION

The contacts and symbols found on ladder rungs are called instructions. Currently there are 107 instructions available for SLC 5/03, 5/04, and 5/05 modular processors with up-to-date operating systems. The SLC 5/01 processor has 52 instructions, while the 5/02 supports 71 instructions. Fixed members of the SLC family have a limited instruction set by their nature. Refer to RSLogix 500 help for additional information regarding the instructions and the specific processors they are used with. This exercise will introduce the basic bit or relay instructions. After the instructions are introduced, we will develop our first ladder program using bit instructions.

The SLC 500 uses the following basic bit or relay logic instructions to represent inputs and outputs. These instructions operate on a single bit of data associated with the bit's assigned address.

Figure 13-1 illustrates SLC 500 and MicroLogix 1000 bit instructions symbology, name, and normal state, along with when each instruction is either true or false.

UNDERSTANDING XIO AND XIC INSTRUCTIONS

During the processor scan, while running the user program, the processor examines the ON/OFF states of data file bits during the input update portion of the scan. After all inputs have been read and the input image table updated, the processor solves the ladder logic. During the program scan, one rung is scanned at a time by evaluating instructions starting from the left-most on the current rung and working right in search of true instructions which will provide logical continuity. Logical continuity makes a rung true. As a result of a logically true rung, outputs are turned on during the output update portion of the scan.

RSLOGIX 500 BIT INSTRUCTIONS				
Symbol	Name	Mnemonic	Normal State	Operation
—| |—	Examine If Closed	XIC	Open	Examines the addressed data table bit for a 1 to be true and a 0 to be false.
—|/|—	Examine If Open	XIO	Closed	Examines the addressed data table bit for a 0 to be true and a 1 to be false.
—( )—	Output Energize	OTE	Off	OTE instruction will set a bit in the addressed data table bit location if a true path of input instructions exist.
—(L)—	Output Latch	OTL		Address bit will be latched on when input conditions are true for a minimum of one processor scan.
—(U)—	Output Unlatch	OTU		Address bit will be unlatched on when input conditions are true for a minimum of one processor scan.

Figure 13-1   The basic RSLogix 500 data files.

To better understand logical continuity, let's compare it to electrical continuity. Electrical continuity is a complete path for electron flow through a circuit so as to turn on the controlled device. Figure 13-2 illustrates that for this conventional circuit to turn on Light #1, we must have current flow through Switch #1 and Switch #2. We now have a complete circuit, or electrical continuity, which will allow Light #1 to light.

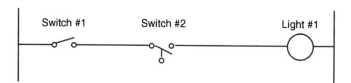

Figure 13-2   Electrical continuity needed to provide complete circuit.

Logical continuity in PLC ladder rungs operates in a fashion similar to electrical continuity. PLC ladder rungs, such as those illustrated in Figure 13-3, do not have electrical continuity as in a conventional circuit. The PLC ladder rung illustrated in Figure 13-3 must have both instructions true before the processor will send an output signal to the output module. If inputs I:1 and I:2 are true, output L1 is true.

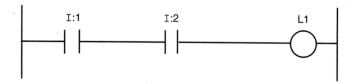

Figure 13-3   PLC ladder rungs have logical continuity, not electrical continuity.

Remember, input signals are placed in the appropriate data file during the input portion of the scan. When the logic is solved during the program scan portion of the processor scan, the processor is looking for logical continuity between Switch #1 and Switch #2. Each instruction is evaluated as either true or false. A true instruction is represented as a logical 1, whereas a false instruction is represented by a logical 0. These 1s and 0s are placed in memory as a logical representation of the true or false state of each instruction. Logical continuity simply means that there is a continuous flow of "true" instructions from the left power rail to the output instruction. The processor then turns on or off the physical output devices as directed by the solved rung logical status placed in the output status file. Figure 13-4 illustrates the truth table for Figure 13-3's PLC ladder rung.

TRUTH TABLE FOR FIGURE 13-3 LOGIC		
Inputs		Outputs
XIC	XIC	OTE
False (0)	False (0)	False (0)
False (0)	True (1)	False (0)
True (1)	False (0)	False (0)
True (1)	True (1)	True (1)

Figure 13-4   Truth table for Figure 13-3.

Fill in the table in Figure 13-5 with information for the basic bit instructions. For additional information, refer to your text or RSLogix help.

BASIC BIT INSTRUCTIONS DATA					
	XIO	XIC	OTL	OTU	OTE
Symbol	-/-	-/-	-(L)-	-(U)-	-( )-
Explanation					
When logic 0	True	False	False	False	False
When logic 1	False	True	True	true	True
Input or output?	Input	Input	output	output	output

Figure 13-5   Table of basic bit instructions.

CONTROLLING BITS OF LOGIC IN A LADDER PROGRAM

A ladder program consists of a number of individual rungs containing one or more input instructions and usually one (however, there may be multiple) output instruction. The following exercise refers to the following rungs of logic (Figure 13-6).

THE LAB

For this lab exercise we will develop the ladder program illustrated in Figure 13-6.

NOTE: As we enter input and output instructions on ladder rungs beginning with this lab programming exercise, the following addressing format will be used in the lab manual steps I:1/2 (I:0/2). The first address, I:1/2, in this example will be used for a modular PLC. The address in parentheses (I:0/2) will be used if you are using a fixed PLC.

1. _____ Open the Begin Project.
2. _____ Make sure the ladder window is active.
3. _____ For this exercise we will be selecting our programming instructions out of the Tabbed Instruction toolbar. The Tabbed Instruction toolbar is illustrated below with the parts we are going to use in this exercise identified. Version 4.0 of the RSLogix 500 software reintroduced customization of the User tab. Your tab may look slightly different from that in Figure 13-7.
   A. Click on the User tab to activate the tab and the associated instructions.
   B. Click here to add a new rung to your ladder.
   C. Click here to add a branch.
   D. Examine-if-closed instruction.
   E. Examine-if-open instruction.

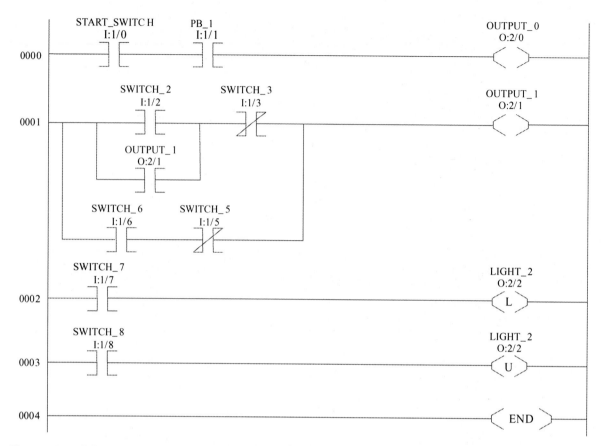

Figure 13-6 RSLogix ladder program.

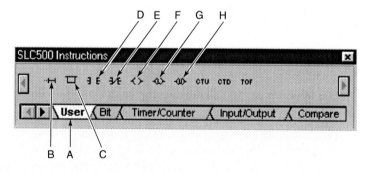

Figure 13-7 RSLogix 500 Tabbed Instruction Toolbar.

    F. Output energize instruction.

    G. Output latch instruction.

    H. Output unlatch instruction.

4. _____ Your ladder currently has one rung identified as rung 0000. This rung also has the word END in an oval on the right. This is the last rung of your program. The END identifies this as the last rung to the processor.

5. _____ Click on the New Rung icon 3 times on the tabbed instruction toolbar.

6. _____ Three new rungs will be added to your ladder.

7. _____ The new rungs will be identified as rungs 0000 through 0002, while the END rung is the last rung.

8. _____ Before we begin programming instructions on ladder rungs we want to turn off the Address Wizard. The Address Wizard was introduced to the RSLogix 500 software with version 4.0. The Wizard is a helpful tool for assisting the programmer in selecting the next unused address as instructions are placed on the ladder rung. If you are using RSLogix 500 version 4.0 or later, complete steps 8 through 13. If using an older version of the software, skip to step 15. We will work with the Address Wizard in a later lesson.

9. _____ From the Windows menu bar, click on Tools.

10. _____ Select Options.

11. _____ From the Systems Options dialog box, select the Xref/Address Wizard tab.

12. _____ Uncheck Enable Address Wizard.

13. _____ Click Apply.

14. _____ Click OK to leave this dialog box.

15. _____ Click on the area to the right of the first rung that identifies it as rung 0000. This will select the rung.

16. _____ Click on the Examine-if-closed instruction from the Tabbed Instruction toolbar.

17. _____ The instruction should appear on the ladder rung. Refer to Figure 13-8 for RSLogix ladder references.

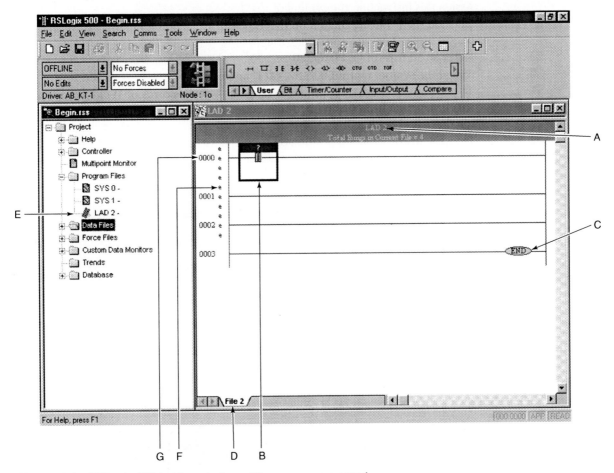

Figure 13-8   RSLogix 500 ladder window with new rungs created.

A. Identifies ladder file displayed.

B. XIC instruction just programmed awaiting address.

C. Ladder program END of program rung.

D. Ladder window tab identifying open ladder windows.

E. Ladder file 2 in Program Files folder within the Project Tree.

F. The "e" identifies the rungs that are under edit.

G. Rung number 0000.

18. _____ To enter an address for the XIC instruction, double-click on the "?" above the instruction to open the text box where you enter the address. Refer to (B) in Figure 13-8.

19. _____ If you are using a modular SLC 500 PLC, type in the address I:1/0 (I:0/0).

20. _____ Press Enter.

21. _____ Right-click on the instruction to enter a symbol.

22. _____ From the menu, select the edit symbol.

23. _____ Type in START SWITCH.

24. _____ Press Enter. The symbol START_SWITCH should be directly above the address. The symbol is attached to this address anytime it is used again in this project.

25. _____ Click on the XIC instruction on the Tabbed Instruction toolbar to add an instruction in series with the first.

26. _____ Enter a valid instruction address.

27. _____ Assign this instruction the symbol of PB_1.

28. _____ Click on the OTE instruction on the Tabbed Instruction toolbar. This will place an output instruction on your ladder rung.

29. _____ Enter an address of O:2/0 (O:0/0).

30. _____ Assign the symbol Output 0 to the OTE instruction.

31. _____ To program the next rung using the drag-and-drop features of Windows, select the XIC instruction from the Tabbed Instruction toolbar by pressing the left mouse button down and holding it.

32. _____ Drag the instruction into position on rung 1, illustrated as (A) in Figure 13-9. Notice the small boxes that appear on the ladder program. See Figure 13-9. These are targets that will accept the instruction as you drag and drop it into the desired position.

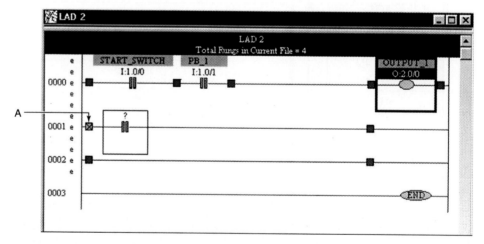

Figure 13-9   RSLogix 500 ladder showing drag-and-drop targets.

Keep the following points in mind as you drag and drop the instruction into place:

A. One of the target boxes must display an "x" before the instruction will be accepted in that position.

B. You can drop the instruction at any target position as long at it is displaying an x.

C. If you attempt to drop an instruction where no target exists, or if the target has not displayed an x because you are not close enough to it, the instruction will disappear.

33. _____ Double-click on the "?" above the instruction and enter the address I:1/2 (I:0/2). Add the symbol SWITCH_2.

34. _____ Drag and drop an XIO instruction in series with the previous instruction. Address this instruction as I:1/3 (I:0/3). Include a symbol of SWITCH_3.

35. _____ Place an OTE instruction at the rung's output with the address O:2/1 (O:0/1). Enter the symbol as illustrated in Figure 13-6.

Next we will program the parallel branch around inputs I:1/2 and I:1/3.

36. _____ Select the rung branch from the Tabbed Instruction toolbar (refer to Figure 13-7) by holding down the left mouse button. Drag the branch and drop it into position as illustrated in Figure 13-10.

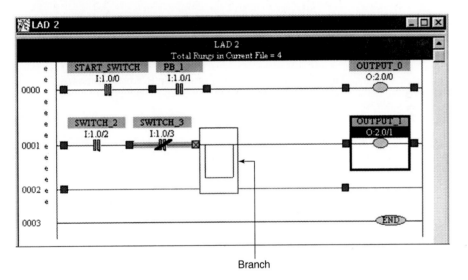

Branch

Figure 13-10 Positioning the rung branch.

37. _____ To position the branch, select the cursor by holding down the left mouse key on the cursor, as shown in (A) Figure 13-11. Drag the cursor into position as illustrated by (B). Release the mouse button when the target displays an x to accept the branch.

38. _____ Program two XIC instructions on the parallel branch as illustrated in Figure 13-6. Address the instructions as I:1/4 (I:0/4) and I:1/5 (O:0/5).

39. _____ Assign symbols to the instructions as illustrated.

40. _____ To extend the branch down and add another branch, click on the lower left-hand corner of the branch to place the cursor at that point. Refer to (C) in Figure 13-11.

41. _____ Right-click and click on Extend Branch Down from the menu. An empty branch should be in place.

42. _____ Add the XIC instruction as illustrated.

43. _____ A nested branch is a branch within a branch. The branch around I:1/2 on rung 0001 is a nested branch. To insert this branch, click (A) in Figure 13-12.

44. _____ Select the rung branch from the Tabbed Instruction toolbar by holding the left button on the mouse.

45. _____ Drag and drop the nested branch into position as illustrated by (B).

46. _____ Click and drag the highlighted right branch leg to the right of the XIC instruction. When the target displays an x, release the mouse button. There should be an empty branch around the SWITCH_2 instruction.

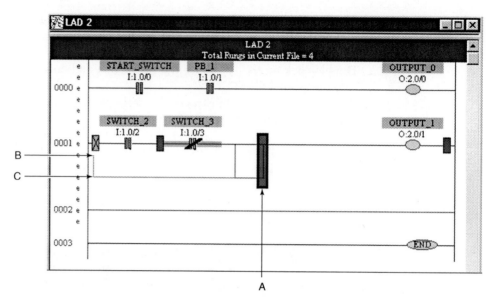

Figure 13-11 Inserting a branch in RSLogix software.

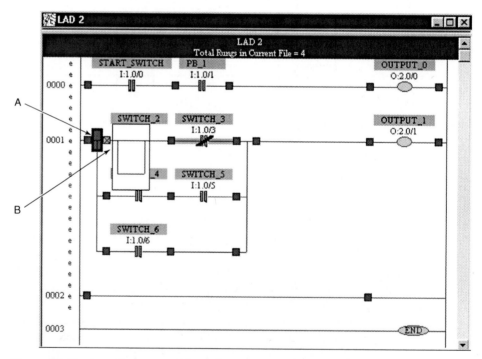

Figure 13-12 Inserting a nested branch.

47. _____ Drag and drop an XIC instruction onto the empty nested branch.

48. _____ Double-click on the "?" in the address area of the instruction. Instead of typing in an address, type in the symbol for OUTPUT_1. Press the Enter key.

49. _____ Create rung 0002 with an XIC as illustrated in Figure 13-6.

50. _____ Program the rung's output instruction as an OTL (L) instruction from the Tabbed Instruction toolbar. Enter the address O:2/2 (O:0/2) and symbol as shown.

51. _____ Program rung 0003. Program the output instruction as an OTU (U) instruction from the toolbar. Remember, the OTL and OTU instructions are often used in pairs.

Their addresses and symbols will be the same. Since the address and symbol are the same as the OTL instruction above, it is easy to drag and drop the address from the above instruction. Left-click and drag the address from the OTL instruction to the target above the OTU instruction.

52. _____ Select the OTU instruction by left-clicking on it.

Assume you are not familiar with the (U) instruction. The easiest way to get more information on an unfamiliar instruction is to select the instruction and press the F1 key on your computer keyboard.

53. _____ After pressing the F1 key, the Help screen for the OUT instruction should display. Review the Help screen contents to familiarize yourself with not only additional information on the instruction, but also with the format of information available regarding instruction help. This same procedure can be used to obtain help for any ladder instruction.

54. _____ Insert a new rung and program an XIC and OTE instruction without any addresses.

Verification

Before a project can be downloaded it must be verified for correctness. Any errors must be corrected before downloading. When downloading, verification is accomplished in the background. If the project is free of errors, the project will be downloaded. When developing or editing a project, the project can be manually verified by using one of two options: Verify File or Verify Project. To verify the current ladder file displayed in the ladder window, select the Verify File icon from the Windows toolbar (A) in Figure 13-13. To verify the entire project and all ladder files, select the Verify Project icon (B) in Figure 13-13.

55. _____ To verify the file, click on the Verify File icon (A).

56. _____ The Verify Results window opens and displays the errors on the Verify Results tab.

57. _____ Refer to (C) in the figure to identify the first error in ladder file 2, rung 4, instruction 1. Clicking here will move the cursor on your ladder program to the instruction with the error. Refer to (D) in the figure. The question mark above the instruction means there has been no address programmed; this is an error.

58. _____ Refer to (E) in Figure 13-13 to identify the second error in ladder file 2, rung 4, instruction 2. Clicking here will move the cursor on your ladder program to the instruction with the error (F). Click here to select the instruction in error. This should be the OTE instruction on rung 4. There is no address assigned.

59. _____ Let's assume this rung was inserted into our program by error. Right-click on the OTE instruction.

60. _____ Select Delete. The OTE instruction should be deleted. This is an example of how an instruction can be deleted.

61. _____ Right-click on the rung number 0004.

62. _____ Click on Delete from the menu. The rung will be deleted.

63. _____ Assume we really did not wish to delete the rung. Click on the Undo icon, (G) in Figure 13-13. Each time it is clicked it will undo the previous action. As you click on the Undo icon, see the deletions return to the ladder rung.

64. _____ After you have finished experimenting with the Undo icon and its operation, delete rung 0004.

As long as we are experimenting with editing, let's look at some other editing features before we download our project and run the project on our PLC.

65. _____ Right-click on the Switch 5 instruction.

66. _____ Click on Change Instruction Type from the menu.

67. _____ Notice the instruction mnemonic XIC is highlighted.

68. _____ Change the mnemonic to XIO.

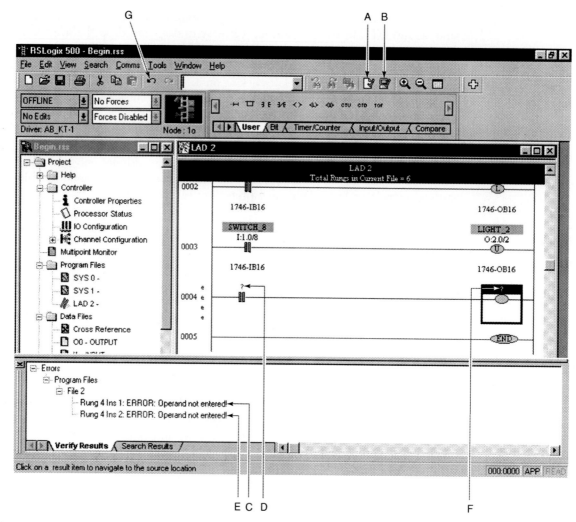

Figure 13-13 Verification of ladder to check for errors.

69. _____ Press Enter. The instruction should have changed to an XIO. This is an easy way to change or edit any instruction on your ladder.

70. _____ Double-click on the address I:1/4. The address should be highlighted.

71. _____ Press the left arrow key on your computer keyboard. This should un-intensify the address.

72. _____ Press the backspace key on your keyboard to delete the bit number from the address.

73. _____ Type in a 6 for the bit number and press Enter. This is an example of editing an address.

74. _____ Click in the lower left corner of the branch, to the left of Switch 6 on the bottom rung of the branch, to select the branch rung.

75. _____ Delete this branch rung.

76. _____ Click on the Verify File icon again. The project should verify. The message "Verify has completed, no errors found" should display in the lower left corner of the computer screen.

77. _____ Save your project as Ladder 1.

78. _____ Select download from the on-line toolbar.

79. _____ The revision note will display. Enter a note if you wish and click OK.
80. _____ View the messages displayed and respond appropriately.
81. _____ In response to the question, "Do you wish to go on-line?" click Yes.
82. _____ From the on-line toolbar, select run. Put the processor in run mode.

As the program runs, answer the following questions.

1. The START_SWITCH is in series with PB_1. If you press the START_SWITCH what must you do to PB_1 to make output 0 true? _____

   What kind of logic is this? _____

2. Explain logical continuity and how it pertains to rung 0000. _____

   _____

   _____

   _____

3. Where are the bits stored representing the addresses on rung 0000? _____

   _____

4. When SWITCH_2 is pressed and released, explain how the two instructions referencing OUTPUT_1 react. _____

   _____

   _____

   _____

   _____

5. After SWITCH_2 is pressed and released, explain the function of SWITCH_3. _____

   _____

   _____

   _____

   _____

6. Explain the function of the (L) instruction on rung 0002. _____

   _____

   _____

   _____

   _____

7. Is the instruction referred to in question 6 retentive or nonretentive? _____

   What does this mean? _____

   _____

   _____

   _____

   _____

8. Press and release SWITCH_7. Explain what you observe. _____

_____

_____

_____

_____

9. What is the function of SWITCH_8? _____

_____

_____

_____

_____

10. Explain the function of the (U) instruction. _____

_____

_____

_____

_____

Lab B: Using Internal Bits to Make a Push-On/Push-Off Push Button

OBJECTIVES

Upon completion of this laboratory exercise, you should be able to:

- understand Bit File addressing
- create a new Bit File
- create PLC ladder logic using bit instructions

INTRODUCTION

For this exercise we need to interface a single momentary push button so that when an operator presses the button once the process starts, and when the button is pushed a second time, the process stops. The exercise will look into developing the ladder program to make this push button work as dictated by the application.

Developing the ladder logic to accomplish this will include a number of instructions needed to set up the proper operating sequence, which will not be physical, outside input points or output points. These instructions will be assigned internal memory addresses in a file called the bit file. The bit file stores single-bit addresses for storage of the ON or OFF state of the instructions associated with the assigned address. When real-world references are not needed, but instructions like XIO, XIC, OSR, OTL, OTU, and OTE need to set up programmed logic which will only be used internally to control other logic, internal bits and the bit file are used. Some manufacturers call these bit references; others refer to them as internal coils, internal relays, or internal bits. In most cases these names are synonymous.

When the operator presses the push button, the PLC needs to see an input pulse for just one scan. If, as a result of contact bounce, multiple input pulses are seen by the PLC, these multiple signals could be interpreted as multiple start or stop signals, and possibly cause unpredictable operation.

To ensure that only one input signal is input to the PLC, the momentary push button's input instruction will be programmed in series with a one-shot instruction. We will develop internal logic, or instructions, to accomplish the task at hand. The internal logic we will develop will establish how the push button's input signal is to be handled by the PLC. Internal logic will consist of XIC and XIO instructions that have internal memory addresses rather than real-world input or output addresses. Internal memory addresses are stored in a file similar to the input status or output status data file. Whereas output status data was stored in file zero, input status data stored in file one, and processor status data stored in file two, internal bit data will be stored in data file three, the bit file. Bit file three is the default bit file created when your processor creates a processor file. Any unused data file above file eight can be configured as an additional bit file.

We will need to add these internal instructions between the push button's input instruction and the real-world output device's output instruction. Bit file three is typically used for internal bits, shift registers, and sequencers. A bit file is made up of 255 one-word elements. One element is a word containing sixteen bits. Each bit is individually addressable. There are a total of 4,096 bits (bits 0 through 4,095) that make up a bit file. Figure 13-14 illustrates an example of a bit file.

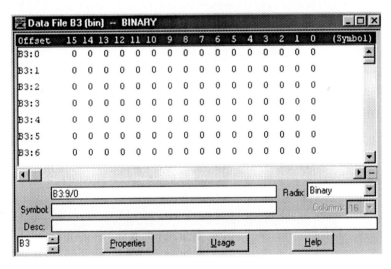

Figure 13-14   RSLogix 500 Binary or Bit file.

EQUIPMENT REQUIRED

One normally open push button is to be wired to PLC input I:1/1 if you have a modular PLC. Connect the push button to input address I:0/1 if you have a fixed SLC 500 or a MicroLogix 1000.

BIT FILE ADDRESSING FORMAT

The bit file addressing format is very similar to input status file and output status file addressing. A typical bit file address is illustrated in Figure 13-15.

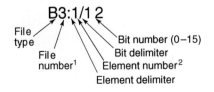

Figure 13-15   Addressing for bit file 3, word or element 1, bit 12.

Alternate Bit Addressing Format

Bit file bits can also be addressed by identifying the file type, file number, and bit number. Since there are 4,096 bits in a bit file and the first bit is 0, the bit range is from 0 to 4,095. Addressing by bit number format is illustrated in Figure 13-16.

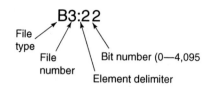

Figure 13-16 Bit address for bit file 3, bit 22.

The example address from Figure 13-16, B3:22, can easily be converted to the element-and-bit addressing format by first determining how many 16-bit words are used. Divide the number of bits by 16 (bits per element) to get the number of elements. Add one to the number of elements to include the element containing the remainder bits. This will give you the total number of elements, which is also going to be the element number in your address.

The element containing the remainder bits should contain fewer than 16 bits. Determine the bit position in the remaining element to get the bit number in your address. Remember to count bit 0 in this element. Two examples are illustrated below:

Example One: Determine element and bit address for B3:22.

1. Bit Number is 22; 22 divided by 16 equals 1 with a remainder of 6. One element, element zero, plus the element containing the remainder bits, puts the element address at B3:element 1.
2. The six remainder bits are found in element 1. Counting from bit zero, the seventh bit in element one is bit 6.
3. The converted address would be B3:1/6, as illustrated in (A) Figure 13-17.

Example Two: Determine element and bit address for B3:100.

1. Bit Number is 100; 100 divided by 16 equals 6 with a remainder of 4. From element zero, the sixth element in our bit file is element five. Adding the element containing the remainder bits, the element address portion of the converted address is B3:element 6.
2. The four remainder bits are found in element 6. From bit 0, the fifth bit in element six is bit 4.
3. The converted address would be B3:6/3, as illustrated in (B) Figure 13-17. An alternate address would be B3:100.

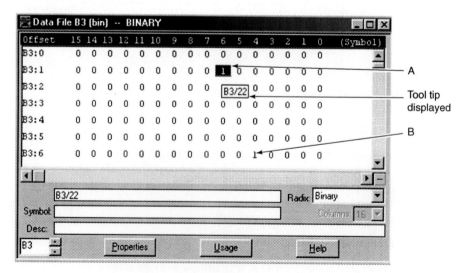

Figure 13-17 Bit file address for B3:22 bit file position.

CREATING AN ADDITIONAL BIT FILE

Even though bit file 3 is the default, automatically created when a new processor file is created, additional bit files may be created for additional, or separate, bit storage. This exercise will show you how to create an additional bit file, bit file 9. We will then create a ladder program containing bit instructions.

THE LAB

Creating a New Binary File

NOTE: If you are using a fixed PLC, such as a MicroLogix 1000, the data files are not expandable. The binary file is limited to 32 elements in file 3 only. Skip ahead to the programming exercise and develop the same program using binary file 3.

1. _____ Right-click on the Data Files folder (A) as illustrated in Figure 13-18.

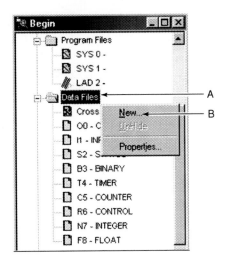

Figure 13-18   Click on New to create a new data file.

2. _____ Click on New (B) from the menu. The Create Data File dialog box should open as illustrated in Figure 13-19.

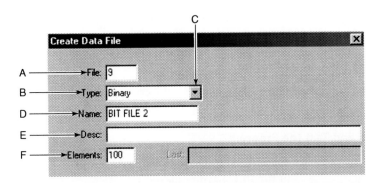

Figure 13-19   The top half of the Create Data File dialog box.

3. _____ The file we wish to create is File 9. Enter the file number at (A) in Figure 13-19.
      This dialog box will display the next available file that can be created. In this situation, File 9 is the next available file.

4. _____ The file type is identified in (B). Click on (C) to display the selections. For this exercise select Binary.

5. _____ Type in the file Name as illustrated by (D).

6. _____ Type in a file Description, if you wish, at (E).

7. _____ Enter the number of Elements the file is to contain at (F).

   The total number of elements for a modular SLC 500 processor is 256. The elements will be numbered 0 to 255. At this point there are two options. If you know how many elements will be needed for this project, type them in here. This value can be modified at any time by simply returning to this dialog box and typing in a new number. If the number of elements is not known, leave the number of elements at the default, 1. As you add elements while programming off-line, the file will be expanded automatically.

8. _____ When completed, click on OK at the bottom of this dialog box to close it. The new binary file, B9, should display under the data files folder. Refer to Figure 13-20.

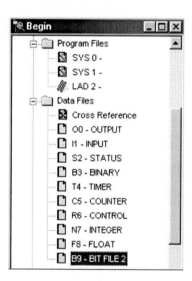

Figure 13-20   Newly created data file, bit file B9.

9. _____ Right-click on the newly created binary file, B9. Notice some of the selections available.

10. _____ Click on Rename. Experiment by giving the file a new name.

11. _____ When completed press Enter.

Developing the Ladder Program

12. _____ Figure 13-21 contains the ladder program for this section. Use your programming skills to create the program as shown. Don't forget to use binary file 9.

Download and Test the Program

1. _____ Save and go on-line to download your program to your PLC.

2. _____ Put the PLC in run mode.

3. _____ As your program runs, verify correct operation after you understand how the program operates.

4. _____ Go off-line and add rung and instructions comments to rungs 1 and 2 describing their operation.

5. _____ When completed with the program, save the program to your student floppy disk.

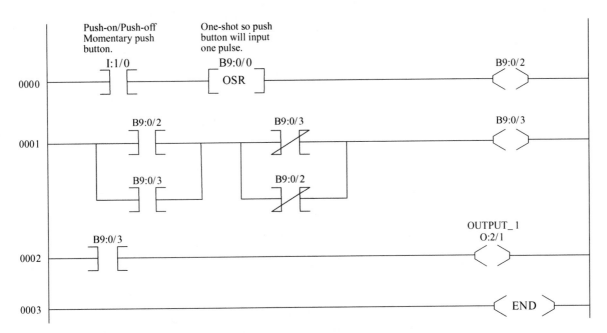

Figure 13-21   Ladder program for the push-on and push-off programming exercise.

REVIEW QUESTIONS

1. How many binary files are allowed in a MicroLogix 1000 PLC? _____

2. In the MicroLogix 1000 PLC, how many elements can make up the binary file(s)? _____

3. How many binary files are allowed in a modular SLC 500 processor like a 5/04? _____

4. In a modular SLC processor, how many elements can make up the binary file(s)? _____

5. User-defined files can be created starting with file 9 in modular processors. What is a user-defined file? _____

_____

6. There are additional user-defined data files other than the ones we have worked with. Where would you look to find information on the available user-defined data files? _____

_____

_____

_____

_____

7. List the steps taken to find the data requested in question 6.

1. _____

2. _____

3. _____

4. _____

5. _____

6. _____

7. _____

8. _____

8. If you wanted to print this help topic, how would you proceed from this point? _____

_____

_____

_____

9. This exercise used bit file bits to control the output. Why did we use these internal bits, internal coils, or internal relays as they are sometimes called? _____

_____

_____

_____

10. What options are available when you right-click on a data file? _____

_____

_____

_____

_____

Lab C: Understanding Relay Instructions

OBJECTIVES

Upon completion of this laboratory exercise, you should be able to:

- select proper PLC instructions
- develop ladder logic from functional specifications
- interpret ladder rungs to determine when they are true or false
- verify correct programming of a normally open start push button and a normally closed stop push button

INTRODUCTION

This exercise will give you practice selecting the proper instruction, or practice creating ladder rungs, for a specific specification. Complete each question regarding proper instruction selection or ladder rung creation.

THE LAB

1. If input bit I:1/7 is normally a *one* and the rung needs to be true when the input status table bit is a *zero*, what instruction should be programmed on the PLC ladder rung? _____

2. Input bit I:1/2 is a *one*, input bit I:1/8 is a *one*, the rung is to be true when both input status table bits are *ones*.

    A. Draw this PLC ladder rung. Use O:2/0 as your output.

    B. What instructions will be programmed on your PLC ladder rung? _____
    _____
    _____
    _____

    C. What happens if the I:1/8 input status bit goes false? _____
    _____

3. If input bit I:1/2 is a one, input bit I:1/8 is a one; develop a rung that will be true when either of these input status table bits is a one.

    A. Draw the PLC ladder rung. Use output O:2/1 as the output.

    B. What instructions were programmed on the PLC ladder rung? _____
    _____

    C. What happens if I:1/8 status bit goes false? _____
    _____

4. We have a piece of machinery with a table where the part being manufactured is held. The table moves, allowing work on the part to be done by the machine. Currently the table is in position holding a normally open limit switch closed. What PLC instruction should be programmed on the ladder rung so the instruction would be true when the table moved away from the limit switch? _____

_____

5. We have an inductive proximity switch looking at a drill bit. An inductive proximity sensor is used to sense the presence of metal. The proximity switch is a normally open style, so it is closed when the drill bit is seen by the switch. We want the PLC to send an alarm to the operator if the drill bit breaks, or if the cable from the drill bit sensor were to become damaged and no signal was being sent. What instruction would be programmed so the rung becomes true in an alarm condition? (When an alarm condition is detected, the machine shuts down and the alarm bell sounds.) _____

_____

6. A robot is busy assembling our product. Around the cell are clear plastic doors that are to stay closed unless a maintenance person needs to go in and repair the robot. Each door has a switch connected to a PLC input. If any door is accidentally opened, the robot is to stop all movement. If a forklift were to hit the cell and cut the cable, the robot is also to stop all movement. In case of switch failure or damage, the robot must be fail-safe.

A. Should a normally open or a normally closed switch be installed on each of the four doors? _____

_____

B. What instruction should be programmed on your PLC ladder rung? _____

_____

C. Thoroughly explain your choices. _____

_____

_____

_____

7. Figure 13-2 in the text refers to the incorrect conversion of a conventional, start-stop schematic to PLC control. Explain in detail why programming a PLC ladder in this manner will not operate correctly. _____

_____

_____

8. When developing a typical start-stop rung of PLC logic, why should the stop push button be programmed as the first instruction on the rung? _____

_____

9. Explain what is meant when OSHA requires that a stop push button be fail-safe. _____

_____

_____

10. When converting a conventional schematic to PLC control, what is the first step? _____

_____

11. Why do we separate inputs so each input provides a separate input signal as an input?

_____

_____

_____

12. Illustrate a typical 8-point input module and how it would be wired in a circuit with two limit switches in parallel, and then two limit switches in series. Include input addresses in your drawing.

13. A normally open, held closed limit switch is wired into I:1/5. What status bit would be found, and in what position of the 16-bit word representing the input module? _____

_____

_____

_____

_____

14

Motor Starter Hardware Interface

PREREQUISITES

Before attempting this lab exercise you should have a solid knowledge of wiring hardware relays, series and parallel hands-on wiring with 120 volts AC line voltage, and have hooked up a similar motor starter with conventional hard wiring.

Safety First!

1. ———— DANGER—120 VOLTS AC LINE VOLTAGE WILL BE PRESENT!
2. ———— Before beginning, have your instructor review and demonstrate this lab exercise so you know what to do, and what to expect as you proceed.
3. ———— Have your instructor review safety procedures for using your particular lab equipment.
4. ———— To ensure personal safety, have your instructor check your hookup before applying power!
5. ———— Examples and hookup exercises have been simplified for instructional purposes and may not prove acceptable in actual industrial applications. Check and follow all applicable codes and ordinances when installing and working with electrical equipment.

Textbook

Study *Introduction to Programmable Logic Controllers*, 3E text, Chapter 14, "Understanding Relay Instructions" before attempting this hands-on lab. Review texts from your industrial electricity classes on motor starters and wiring.

OBJECTIVES

Upon completion of this laboratory exercise, you should be able to:

- convert a conventional relay ladder diagram to PLC format
- create a PLC ladder and program from this converted ladder diagram
- hook up and interface a start-stop station to the PLC
- hook up a motor starter as an output device from your PLC
- verify correct programming of a normally open start push button and a normally closed push button

INTRODUCTION

We will be using a typical motor starter, along with a start-stop push-button station, to interface to a PLC. Figure 14-1 illustrates the major parts of an Allen-Bradley Bulletin 509 motor starter.

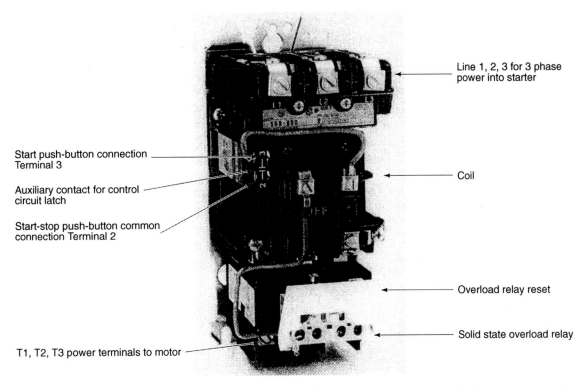

Line 1, 2, 3 for 3 phase power into starter

Start push-button connection Terminal 3

Auxiliary contact for control circuit latch

Start-stop push-button common connection Terminal 2

Coil

Overload relay reset

Solid state overload relay

T1, T2, T3 power terminals to motor

Figure 14-1 Major parts of an Allen-Bradley Bulletin 509 size one motor starter. (Used with permission of Rockwell Automation, Inc.)

Conventional hookup of a motor starter consists of a push-button start-stop station hard-wired to the motor starter. Figure 14-2 illustrates typical conventional hard-wiring of a motor starter. You need to be familiar with hard-wiring a motor starter before you complete this lab.

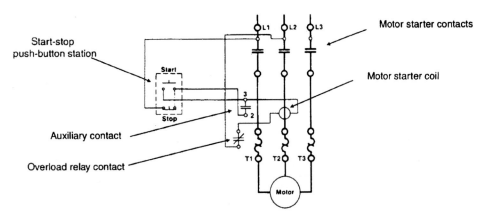

Start-stop push-button station

Auxiliary contact

Overload relay contact

Motor starter contacts

Motor starter coil

Figure 14-2 Conventional motor starter wiring diagram. (Compiled from Rockwell Automation/ Allen-Bradley starter wiring data)

This lab exercise will guide you through the hookup and interface of a motor starter to an SLC 500 programmable logic controller.

EQUIPMENT REQUIRED

To complete this lab exercise you will need:

1. Power, 120 volts AC, necessary tools, and hookup wire.
2. Normally open start, normally closed stop push-button station.
3. A motor starter similar to the Allen-Bradley Bulletin 509-BOD.
4. Allen-Bradley SLC 500 programmable logic controller.

To simplify wiring illustration for this exercise we will configure our PLC with the following two I/O modules:

1746-IA8 input module

1746-OA8 output module

IBM or compatible personal computer

Rockwell Software's SLC 500 programming software

GETTING STARTED

Be sure to check and verify your particular I/O modules and the correct wiring as you work through this exercise. We will start with the conventional motor starter ladder diagram illustrated in Figure 14-3.

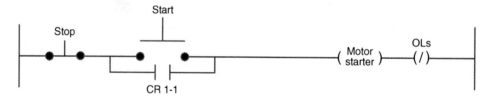

Figure 14-3 Conventional motor starter schematic.

RELAY LADDER CONVERSION TO PLC FORMAT

We need to work on converting our conventional motor starter wiring diagram to PLC format. This will enable us to develop a PLC ladder program and assist us in properly wiring inputs and outputs to our PLC I/O modules.

1. _____ The first step in relay schematic conversion is to determine the inputs and outputs.

   List all inputs: _____

   _____

   List all outputs: _____

   _____

2. _____ The second step is to allocate I/O reference numbers, or addresses. What are the addresses corresponding to each slot of your SLC 500 modular PLC? Figure 14-4 illustrates addresses for the 8-point input and output modules' screw terminals.

	Slot 0	Slot 1	Slot 2	Slot 3
		I:1/0	O:2/0	
		I:1/1	O:2/1	
		I:1/2	O:2/2	
Power supply	5/03 CPU	I:1/3	O:2/3	
		I:1/4	O:2/4	
		I:1/5	O:2/5	
		I:1/6	O:2/6	
		I:1/7	O:2/7	

Figure 14-4   Four-slot SLC 500 illustrating eight-point modules addressed in slots 1 and 2.

3. _____ Assign your input addresses as in the table in Figure 14-5.

INPUT ADDRESS ASSIGNMENT	
List Your Inputs	**Assigned Addresses**
Start normally open push button	I:0/0
Stop normally closed push button	I:0/1
Motor starter normally closed overloads	I:0/2
Motor starter normally open auxiliary	I:0/3

Figure 14-5   Input addresses have been assigned.

4. _____ Assign your output addresses as in Figure 14-6.

OUTPUT ADDRESS ASSIGNMENT	
List Your Outputs	**Assigned Addresses**
Motor starter coil	O:2/0

Figure 14-6   The sole output address has been assigned.

5. _____ Draw the converted conventional relay ladder to PLC format.

6. _____ Figure 14-7 is a conceptional drawing of input wiring for the start-stop push-button station to the PLC input module.

Start-stop stations come in numerous varieties. Some start-stop stations come with only one set of normally open start contacts and one set of normally closed stop contacts. These contacts are built into a small enclosure called a start-stop push-button station. Other start-stop stations may have a removable contact block mounted on each push-button operator. These contact blocks may be single or double circuit. Depending on the particular start-stop station you have, you will have to wire it correctly.

Wiring the push-button station in Figure 14-7: power comes in the left side of Figure 14-7 and is distributed to one side of each push button. The start push button is normally open and will normally send no signal to the input module screw terminal. Notice the stop push button is normally closed. Being normally closed, the stop push button will continuously send an ON signal to the PLC input screw terminal.

7. _____ Figure 14-8 illustrates typical wiring for each of our inputs. There are four separate signals going into, or input into, the PLC input module. The 1746-IA8 module is an Allen-Bradley SLC 500 8-point, 120-volt AC input module. Power from line one goes to each hardware input device and then to the specified input address's screw

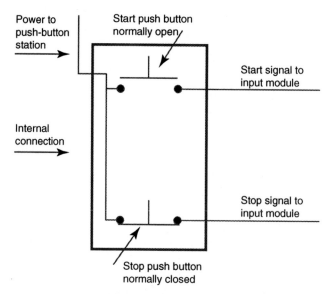

Figure 14-7 Typical start-stop push-button station wiring.

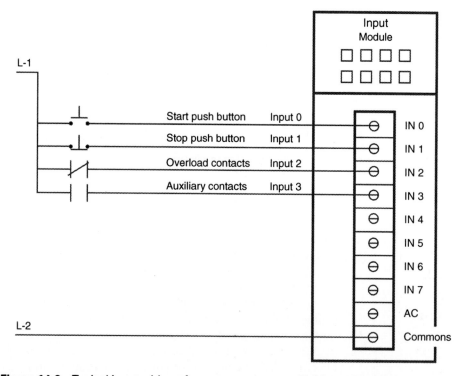

Figure 14-8 Typical input wiring of a motor starter to a PLC input module.

terminal. The internally connected AC common screw terminals complete the circuit to line two. Notice that each input is completely independent as it sends its own signal into the module. Typically, inputs are separated when they are wired to a PLC. This separating of inputs is contrary to conventional hard-wiring you may be accustomed to. Always refer to the manufacturer's wiring instructions before wiring your particular start-stop station.

8. _____ The only output from our PLC is to the motor starter's coil. This output will be wired as illustrated in Figure 14-9, assuming you are using the Allen-Bradley 1746-OA8

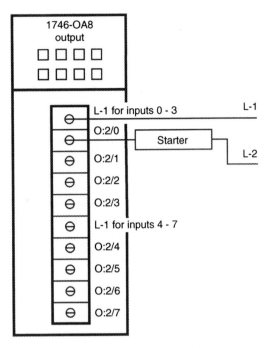

Figure 14-9 Typical wiring of a PLC output module wiring to a motor starter coil.

output module. To wire the module, power from line one goes into the L-1 screw terminal. Think of the output module as a package containing a switch for each output address. When the program finds all rung inputs that are associated with our motor starter coil output as true, a logical *one* will be placed in the output status table, address O:2/0. When the processor updates the outputs, the ON signal in the output status table will be sent to the output module's internal solid-state switch for address O:2/0. An ON signal to O:2/0 module address will close the solid-state switch associated with that output. Power will flow to the starter coil, causing it to energize.

9. _____ Figure 14-10 is a complete wiring diagram for our motor starter interface to the SLC 500 PLC. *This wiring diagram will only be correct if using the exact parts specified in the Equipment Required section.* If using other than specified hardware, see your instructor for modified instructions. Notice that to simplify Figure 14-10, only the top four I/O points are included for the I/O modules.

10. _____ In this lab exercise we are not going to hook up an actual motor to our motor starter. Your instructor can make motor installation an optional portion of the lab exercise. If a motor is to be hooked up, your instructor will supply needed instructions.

THE LAB

In the space provided, check off each step as completed.

1. _____ Select all the parts needed to complete this lab.

2. _____ Place the processor and the correct I/O modules in the proper PLC chassis slots. Slot 3 will not be used. Place any available module in this slot.

3. _____ Do NOT apply power until your instructor has verified correct wiring.

4. _____ Connect a wire from the start push-button screw terminal to PLC input screw terminal I:1/0, as shown in Figure 14-10.

5. _____ Connect a wire from the stop push-button screw terminal to PLC input module screw terminal I:1/1, as illustrated in Figure 14-10.

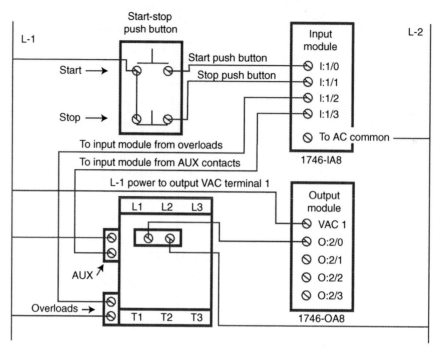

Figure 14-10   Typical motor starter interface to a PLC.

6. _____ Connect a wire from the overload contact screw terminal to the input module screw terminal I:1/2, as illustrated in Figure 14-10.

7. _____ Connect a wire from output module screw terminal O:2/0 to one side of the motor starter coil.

8. _____ Connect a wire from the auxiliary contact screw terminal to the input module screw terminal I:1/3 in Figure 14-10. This normally open contact, I:1/3, is the interlock around the start push button.

9. _____ Connect the L-1 power wires to:

   Start-stop push-button station

   Auxiliary contacts

   Overload contacts

   VAC 1 terminal on your 1746-OA8 output module

10. _____ Connect the L-2 return wires to:

   Input module AC common screw terminal

   One side of the motor starter coil

11. _____ We do not need to connect a motor to verify proper PLC Interface. Consult with your instructor to see if connecting a motor will be added to this application.

12. _____ This should complete wiring your control circuit. Do NOT apply power until completed and your instructor has checked your wiring.

13. _____ Develop your off-line program on the personal computer using your SLC 500 development software. Remember to modify your I/O configuration if necessary.

14. _____ Connect the communications cable between the processor and your personal computer.

15. _____ Apply power to your SLC 500 PLC.

16. _____ Go on-line and download your program into the processor.
17. _____ Put the processor in run mode.
18. _____ When your instructor has verified your control wiring is correct, apply power and test your inputs.
19. _____ When pressing the start button, does the motor starter energize?
20. _____ When you press the stop push button, does the motor starter disengage?

CONSIDERATIONS AS WE COMPLETE THIS APPLICATION

Now that you have successfully completed this lab exercise by hooking up a motor starter and start-stop push-button station to a PLC, let us look at additional considerations for real-world applications.

Suppression of Electromagnetic Interference

Equipment surrounding a PLC may generate considerable amounts of *electromagnetic interference,* or *EMI.* Electromagnetic interference transients, or spikes, result from an inductive load's collapsing magnetic field when the device is switched off. Voltage levels from these spikes can be very high. EMI from noise-generating equipment such as relays, solenoids, and motor starters can cause intermittent problems in PLC operation, as well as damage output module circuitry. To help eliminate EMI transients, isolation transformers should be placed between a PLC and the AC voltage source. Surge suppression devices should also be used on all inductive devices operated by hard contact control devices. Hard contact devices include push buttons, selector switches, and relay contacts. Any inductive output device that is switched by hard contacts, including relay output PLC modules, needs surge suppression. Even if the output control circuit is switched by a solid-state output, and there is a hard contact switching device in the output circuit, surge suppression is necessary. Normally, a solid-state output module directly switching an inductive load may not require surge suppression. However, surge suppression can be added to the inductive load even though there are no hard contacts present in the output circuit, as there is still some degree of protection for the solid-state device. Refer to your hardware manufacturer's instructions on selecting and installing surge suppression on your particular equipment.

Motor starters, typically NEMA size three or four and larger, draw too much current to be controlled by many PLC output modules. In this situation, a switching device needs to be placed between the PLC output module and the motor starter. This device, an *interposing relay,* can be easily switched by the PLC to control the larger sized motor starter.

Interposing Relay

When interfacing a PLC to any motor starter, or inductive load, the load's current draw must be considered in relation to the ratings of the output module. The point where the output module cannot handle the current is where an interposing relay is required in the output control circuit. When the control load is larger than the rating of the selected output module, a standard hardwired control relay is placed between the output module and the load. The output module switches the interposing relay, and the interposing relay switches the load. Care must be taken in selecting the interposing relay. The interposing relay must have inrush and sealed current values within the specifications of the output module. Most control relay contacts are rated for up to ten amps, so verify that the interposing relay contacts can handle the motor starter's coil load. Typically, interposing relays are needed with size three or four and larger motor starters. Figure 14-11 lists the inrush and sealed coils' specifications for various starter sizes.

TYPICAL SEALED AND INRUSH VALUES FOR ALLEN-BRADLEY 3-POLE BULLETIN 509 STARTERS		
	Values Listed in Amps	
Starter Size	Inrush Value	Sealed Value
0	1.67	.25
1	1.67	.25
2	2.09	.25
3	5.74	.39
4	10.62	.60
5	12.96	.835

Figure 14-11 Typical NEMA motor starters' current specifications. (Table data compiled from Rockwell Automation/Allen-Bradley starter wiring data)

Determining the Largest Motor Starter
Our Output Module Will Switch Without an Interposing Relay

The 1746-OA8, a 100–240 VAC triac output module, is rated at one amp continuous current per output point. The maximum surge current per output point for the 1746-OA8 is 10.0 amps for 25 milliseconds. From Figure 14-11, a size three NEMA starter will draw 5.74 amps inrush current, while a size four starter will draw 10.62 amps. Since this module is rated for 10.0 amps maximum surge current, a size four starter will need an interposing relay, even though continuous or sealed current is well below maximum. Each output module chosen for an application should be checked for suitability before using the module.

SUMMARY

This chapter provided the opportunity to get hands-on experience hooking a motor starter to a PLC. You had the opportunity to prove that the normally closed, stop push button must be programmed as a normally open PLC input instruction. Likewise, the normally closed overload contacts must also be programmed as normally open PLC instructions.

Interposing relays are used when an output module cannot switch the current level required to energize and de-energize the motor starter coil. Remember, whenever hard contacts are controlling an inductive load, surge suppression is necessary. Always refer to your hardware documentation when selecting surge suppression for a specific application.

REVIEW QUESTIONS

1. Where would the motor leads have been connected to the motor starter if we had connected a motor to our starter? _____

2. If we had hooked up a motor to our motor starter, where does the power to operate the motor connect? _____

3. What is typically done to obtain a lower control circuit voltage level than the usual 480-volt, three-phase motor voltage? _____

4. Regarding the PLC program rung for controlling the motor starter, explain the extra set of contacts on the parallel branch around the start button. _____

5. Why is the stop push button the first instruction on our PLC ladder rung? _____

_____

6. The normally closed, stop push button is programmed as normally open. Wouldn't it make sense to have the input instruction reflect the actual state of the push button? Explain.

_____

_____

7. What would have happened if you had programmed your stop push button as a normally closed instruction? Explain your answer. Did you try it? _____

_____

_____

8. Why is surge suppression needed on an inductive load? _____

_____

9. Discuss any problems that may be encountered when interfacing larger motor starters to PLCs. _____

_____

_____

10. Do the overload contacts send a *normally open* or *normally closed* signal to the PLC input module under normal running conditions? _____

11. When finished developing your program off-line, you go on-line. Explain what this means.

_____

12. Fill in the table in Figure 14-12 as to whether surge suppression is necessary under the described conditions. (Surge suppression can be added to inductive loads directly switched by solid-state output modules to receive a degree of protection against transient spikes.)

IS SURGE SUPPRESSION NECESSARY?			
Output Module Type	Hard Contact Hardware in Output Circuit?	Field Device	Suppression Needed?
Solid state	No	Motor starter	
Solid state	Yes, Selector switch	Transformer pilot light	
Relay output	Yes, Push button	Transformer pilot light	
Solid state	No	Interposing relay	
Relay output	No	Interposing relay	
Solid state	Yes, Selector switch	Solenoid	
Solid state	Yes, Push button	Pilot light	
Relay output	No	Solenoid	

Figure 14-12 Surge suppression determination.

15

PLC System Documentation

OBJECTIVES

Upon completion of this laboratory exercise, you should be able to:

- program rung comments and page titles
- enter symbols when programming instructions
- add instruction descriptions
- program address descriptions
- identify rung numbers

LAB ONE

The ladder diagram printout, Figure 15-1, has parts identified with numbers. For the following exercise, match those numbers with the term that best identifies the portion of the ladder diagram identified.

Number 1 is:	_____	A. Left power rail
Number 2 is:	_____	B. Right power rail
Number 3 is:	_____	C. Examine if open contact
Number 4 is:	_____	D. Examine if closed contact
Number 5 is:	_____	E. Output energize instruction
Number 6 is:	_____	F. Rung comment
Number 7 is:	_____	G. Instruction comment
Number 8 is:	_____	H. Bit address
Number 9 is:	_____	I. File number
Number 10 is:	_____	J. Rung number
Number 11 is:	_____	K. Input module identification
Number 12 is:	_____	L. Output module identification
Number 13 is:	_____	M. End of program rung
Number 14 is:	_____	N. Input instruction address
Number 15 is:	_____	O. Output instruction address
Number 16 is:	_____	P. Page title

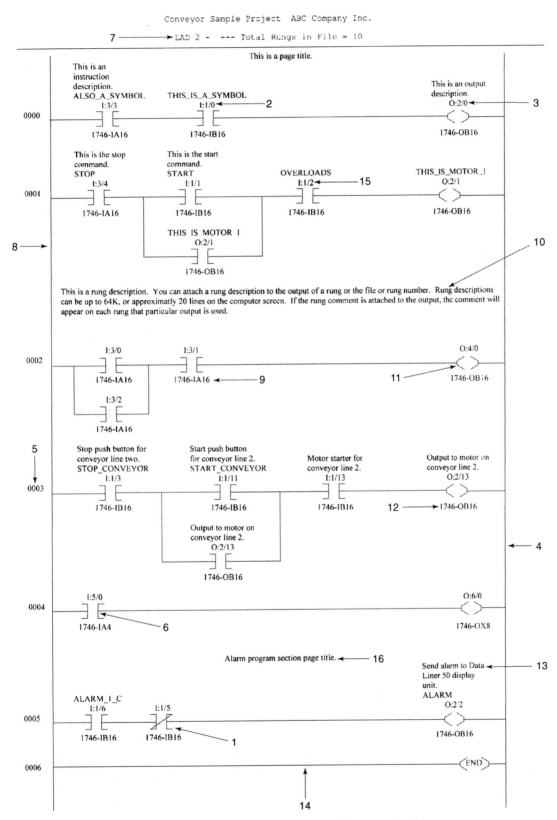

Figure 15-1 Ladder program printout from Rockwell Software's RSLogix 500 software.

LAB TWO

This lab exercise will provide you with practice programming page titles, rung comments, symbols, instruction descriptions, and instruction comments.

Develop the ladder program as illustrated in Figure 15-1. Program all page titles, rung comments, symbols, instruction descriptions, and instruction comments. Refer to the procedures below for the steps to adding each type of documentation.

Adding a Page Title or Rung Comment

A page title is only one line of up to 80 characters. Typically, page titles are not associated with each rung. A page title has two functions. First, page titles can be used to separate a ladder program into logical groups of rungs. As an example, use a page title to group the filling ladder logic, the capping, labeling, packing, or case sealing logic. By grouping ladder rungs using page titles, you can search for page titles using Advanced Diagnostics under the Search function. Secondly, when printing a hard copy of your program, each page title will start a new piece of paper. This way a ladder program can be laid out in logical sections or easily placed in a loose-leaf notebook.

Rung comments can be added to each rung of your ladder program. Rung comments can have up to 64 Kbytes per rung with up to 500 lines displayed. Rung comments are used to add extensive documentation as to the purpose, operation, and in some cases maintenance and troubleshooting for each rung.

Steps to Programming Page Titles and Rung Comments

1. _____ To add a page title or rung comment, right-click on the rung number to which you wish to add the page title or rung comment. The Shortcut menu will display.

2. _____ From the Shortcut menu, select Edit Title (A) or Edit Comment (B) as illustrated in Figure 15-2.

   The text box opens as illustrated in Figure 15-3. If you have version 4.0 or later, you could choose to enter a Rung Comment (A) or a Page Title (B), as a rung comment or page title can be entered in the same text box. If your software is older than version 4.0, you will have to do one at a time from the Shortcut menu. Figure 15-3 illustrates the single text box for version 4.0 or newer where either page titles or rung comments can be entered.

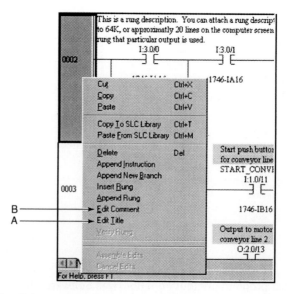

Figure 15-2   Right-click menu.

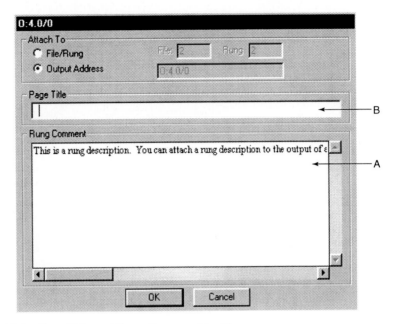

Figure 15-3   Page Title and Rung Comment text box.

3. _____ At the top of the text box is the Attach To area. Select either File/Rung or Output Address to attach the rung comments.

4. _____ Click OK to accept your page titles and rung comments and to return to your ladder view.

5. _____ Remember that the Properties Screen Comment Display tab setup will determine what documentation will be displayed and how it will be displayed. If your documentation does not display on your ladder view, check the Properties setup.

NOTE: If you want to edit a page title or rung comment, simply double-click on the item and the text box will open.

Entering a Symbol

A symbol is used as an identifying label that can be a substitute for an address. Rather than remember an address, such as I:3.0/12, a symbol like Switch 1 or Stop Conveyor 2 can be used instead of the address or in conjunction with the address. All instructions having the same address will have the same symbol. Once the symbol and address are married together, additional programming instances of the address can simply be keyed in using the symbol. Symbols can be up to twenty characters. All letters will be displayed as capitals. Spaces between letters will be designated by the underscore character which will be automatically inserted when you press the space bar. Valid characters are letters A through Z and numbers 0 through 9. Remember, ladder properties can be configured to display the address and symbol, or the symbol only.

Entering a Symbol

1. _____ Right-click on the instruction for which you wish to enter the symbol.

2. _____ A shortcut menu should open. Select Edit Symbol. Refer to (A) as illustrated in Figure 15-4.

3. _____ A small window will open above the instruction for which you wish to enter the symbol.

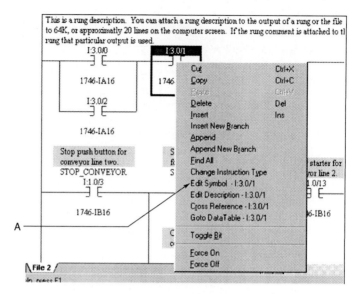

Figure 15-4   Click Edit Symbol to enter the symbol for your instruction.

4. _____ Type in the symbol.
5. _____ When completed, press the Enter key. The symbol should display above the instruction.
6. _____ Remember the ladder Properties Comment Display tab is where you set up if and how symbols are to be displayed. If symbols are to be displayed, the options are Show Symbol Only or Show Symbol & Address.

   To edit a symbol on the ladder rung, simply double-click on the symbol and make your edits. When completed, press Enter.

Address Descriptions

If an additional explanation of a symbol or address is desired, an address description can be placed on your ladder rung directly above the symbol. Address descriptions can be up to five lines and twenty characters per line. The address description will follow any occurrence of the address with which it is associated. Once the address description is entered, any time that particular address is programmed again it will carry the same address description.

Instruction Comment

An instruction comment is similar to an address description. You have a choice of either an address description or an instruction comment. Instruction comments can also be up to five lines and twenty characters per line. An instruction comment differs from an address description in that the instruction comment is tied to an instruction type and its address rather than to an address, no matter what instruction is associated with the address. When you right-click on the instruction and select Edit Description, a dialog box opens as illustrated in Figure 15-5. There are two choices, either Address or Instruction description.

Programming Address Descriptions or Instruction Comments

1. _____ Right-click on the instruction to enter the address description or instruction comment.
2. _____ Select Edit Description from the shortcut menu.
3. _____ Select either Address or Instruction under Edit Description Type.

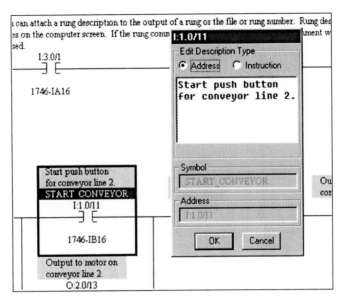

Figure 15-5 Edit Description dialog box.

4. _____ Left-click in the box to enter the text.

5. _____ Type in the description or comment.

6. _____ Click on OK.

7. _____ The description or comment should display above the address and symbol currently associated with the instruction.

8. _____ To edit an existing description or comment, simply double-click on it from your ladder window and edit the text.

NOTE: (1) The properties for displaying your descriptions are configured under Properties, Comment Display tab; and (2) the colors of your address descriptions or instruction comments are set up under Properties, Colors tab.

Figure 15-6 illustrates the difference between an address description and an instruction comment. Address descriptions on rung 0006 are the same even though the instructions are different. The instruction comments on rung 0007 are different. In both cases the addresses are the same, but the instructions are different.

Ladder Program Display Properties

1. _____ Right-click on a blank portion of your ladder window.

2. _____ Left-click on Properties.

3. _____ The tabbed View Properties dialog box opens.

4. _____ Experiment by changing display colors under the Colors tab.

5. _____ Experiment turning documentation on and off from the Comment Display tab.

6. _____ On the Address Display tab, change the bit address format.

7. _____ From the Miscellaneous tab, experiment with rung wrapping, show 3D instructions, show page headers, and show I/O types.

8. _____ Ask your instructor if your completed ladder with documentation needs to be printed out and handed in for credit.

9. _____ When completed, close the RSLogix 500 software.

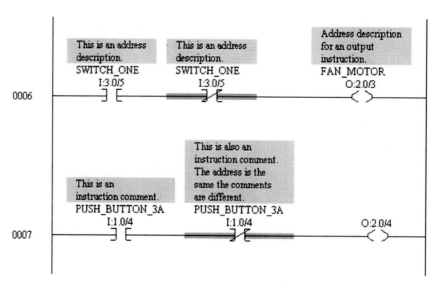

Figure 15-6 Instruction comment and address description differences.

REVIEW QUESTIONS

1. _____ can be added to each rung of your ladder program.

2. Rung comments can have up to _____ per rung with up to _____ lines displayed.

3. A _____ is only one line of up to 80 characters.

4. List the two functions of a page title. _____

5. A _____ is used as an identifying label that can be a substitute for an address.

6. All instructions having the same address will have the same _____.

7. Once the symbol and address are married together, additional programming of the address can simply be keyed in using the _____.

8. Symbols can be up to _____ characters and will be displayed as

   _____

9. Spaces between letters in a symbol will be designated by the _____, which will be automatically inserted when you press the space bar.

10. Valid characters for symbols are _____.

11. If an additional explanation of a symbol or address is desired, an _____ can be placed on your ladder rung directly above the symbol.

12. _____ can be up to five lines and twenty characters per line.

13. The _____ will follow any occurrence of the address with which it is associated.

14. Once the address description is entered, anytime that particular address is programmed again your ladder will carry the same _____.

15. You have a choice of either an address description or an _____.

16. _____ comments can also be up to five lines and twenty characters per line.

17. Explain how an instruction comment differs from a rung description. _____

   _____

   _____

   _____

18. To edit an existing description or comment, simply _____ from your ladder window and edit the text.

19. The properties for displaying your descriptions are configured under the _____

   _____ tab.

20. The colors of your address descriptions or instruction comments are set up under Properties, _____.

16

Timer and Counter Instructions

Lab A: Programming Timers

OBJECTIVES

Upon completion of this laboratory exercise, you should be able to:

- identify the TON, TOF, and RTO timers
- program the TON, TOF, and RTO timers and understand their associated status bits
- download the timer project and go on-line
- create a timer application from a functional specification

INTRODUCTION

The SLC 500 family of PLCs has three timer and two counter instructions. This section will give you an opportunity to work with each of the timer and counter instructions. This lab contains a section on timers and a separate section on counters. The first section will introduce you to the timer instructions.

INSTRUCTOR DEMONSTRATION

Follow along as your instructor demonstrates an RSLogix 500 program containing the timer instructions.

TYPES OF TIMER INSTRUCTIONS

The three types of timers are on-delay, off-delay, and retentive. Figure 16-1 describes the RSLogix timer instructions.

SLC 500 and MicroLogix timers are stored in timer file four in the data file section of the PLC's memory. The default timer file is file four. The timer file is used to store only timer data. Each timer consists of three 16-bit words, called a timer element. There can be many timer files for a single modular processor file. (The MicroLogix 1000 only allows forty timers in timer file four. No additional timer files or timer elements can be created.) Any data file greater than file eight can be assigned as an additional timer file. Each timer file can have up to 256 timer elements.

TIMER INSTRUCTIONS		
Instruction	**Use This Instruction to**	**Functional Description**
On-Delay	Program a time delay before an instruction becomes true.	Use an on-delay timer when an action is to begin at a specified time after the input becomes true. As an example, a certain step in the manufacturing process is to begin 30 seconds after a signal is received from a limit switch. The 30-second delay is the on-delay timer's preset value.
Off-Delay	Program a time delay to begin after rung inputs go false.	An external cooling fan on a motor is to run all the time the motor is running and for five minutes after the motor is turned off. This is a five-minute off-delay timer. The five-minute timing cycle begins when the motor is turned off.
Retentive	Retain accumulated value through power loss, processor mode change, or rung state going from true to false.	Use a retentive timer to track the running time of a motor for maintenance purposes. Each time the motor is turned off, the timer will remember the motor's elapsed running time. Next time the motor is turned on, time will increase from there. To reset this timer, use a reset instruction.
Reset	Resets the accumulated value of a timer or counter.	A reset is typically used to reset a retentive timer's accumulated value to zero.

Figure 16-1   RSLogix 500 timer instructions.

UNDERSTANDING THE TIMER INSTRUCTION

A timer instruction is comprised of the following components. Refer to Figure 16-2 and identify the components of the timers introduced.

A   Mnemonic. The mnemonic is the three-letter abbreviation used to identify the timer instruction.

B   Identifies timer type.

C   The Timer parameter identifies the address in the timer file where the timer and its associated status information is stored.

D   Time Base is the timing interval. SLC 5/03, 5/04, and 5/05 processors have two time base options: .01 and 1 second.

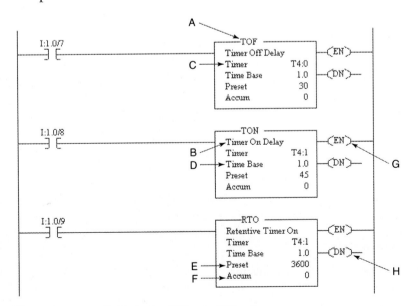

Figure 16-2   Three timers available in the RSLogix 500 software.

E   The preset is an integer value between 0 and +32,767 that identifies how many time base increments the timer is to time before it is considered done.

F   The Accumulated value is our position in the timing cycle. The accumulated value contains the current elapsed time.

G   The EN (enable) is a status indicator that displays whether or not the timer instruction is true, or enabled. The EN is one of the timer's status bits. Status bits will be explained in the next section.

H   DN (done) is also a status indicator displaying when the timer is done timing. The DN is one of the timer's status bits.

TIMER STATUS BITS

In order for the PLC processor and human operator to monitor the current timer status, special bits, called status bits, are used. Timers have three status bits; they are listed in Figure 16-3.

TIMER STATUS BITS			
Status Bit	Identified as	Addressing Example	Functional Description
Timer Enable	EN	T4:0/EN	This bit identifies whether the timer instruction is true or enabled.
Timer Done Bit	DN	T4:0/DN	When the accumulated value programmed for the instruction is equal to the programmed preset value, the Done bit will be true.
Timer Timing	TT	T4:0/TT	The Timer Timing bit is true whenever the timer is actually timing.

Figure 16-3   RSLogix timer status bits.

ADDRESSING SLC 500 TIMERS

Timers are three-word elements. An element is defined as a group of words that work together as a unit. A timer element contains three words that work together to give us a timer instruction. Word zero of the element contains the timer's status bits. Word one contains the preset value, and word two contains the accumulated value. Timer instruction data is in one of two data formats. Status bit data is single-bit data while preset and accumulated value data will be represented as a 16-bit signed integer. Figure 16-4 illustrates the format of the timer element.

Word 0	EN	TT	DN	Reserved Internal Bits
Word 1	Timer Preset Value (PRE). Range 0 to 32,767.			
Word 2	Timer Accumulated Value (ACC).			

Figure 16-4   Timer element data format.

Each timer element has two words containing word data representing the preset and accumulated value. These two words are subelements of the counter element. A subelement is simply a separate piece of an element or instruction that is addressable as a stand-alone unit. The preset value and accumulated value can be addressed individually.

Timer Address Format

Notice there are three formats for addressing timer elements, depending on the part of the element, or one of the subelements, you are addressing. The address format below is to address the entire timer element.

T4:0

If you wish to address a status bit, the timer element is identified followed by a forward slash and either the bit number or bit identifier.

T4:3/DN
T4:3/TT
T4:3/EN

To address a timer subelement such as the preset or accumulated value, the timer element is identified followed by a period and the subelement identifier.

T4:16.PRE
T4:16.ACC

Examples of timer instruction status bit and subelement components are listed in Figure 16-5.

TIMER ADDRESS FORMAT	
Timer Address	**Address Identifies**
T4:0/DN	Timer file 4, element 0, the done bit
T4:46/EN	Timer file 4, element 46, enable bit
T12:9/TT	Timer file 12, element 9's timer timing bit
T21:15.PRE	Timer file 21, element 15's preset value
T4:1/ACC	Timer file 4, element 1, the accumulated value

Figure 16-5   Examples of timer addressing.

TIMER DATA FILE

Timer information is stored in the timer status file. This is one of the data files in the RSLogix software under the Project folder. The default timer file for the RSLogix software is Data File 4. If your processor has enough memory, Data File 4 can have as many as 256 timer elements. The elements are numbered starting with 0 and ending at 255. Data files with file numbers greater than 8 can be used to create additional user files such as timers. Files 8 through 255 can be created to provide additional timer elements. Each additional data file can have up to 256 elements. Keep in mind that most SLC family modular processors support data file expansion, assuming there is enough memory.

Figure 16-6 illustrates the major features of Timer Data File 4 within the Data File folder. The features are identified as follows:

A   Timer Data File 4.
B   Timer T4:1's Done status bit.
C   Enable status bit.
D   Timer element T4:4.
E   Cursor is currently on timer timing bit for timer T4:9/TT (not shown).
F   Enter a symbol here for T4:9.
G   Enter a description here for timer T4:9.
H   Identifies that Data File T4 is currently displayed. Use up and down arrows to move between data files.
I   Click here to view timer file Properties.
J   Click here to view timer file Usage.
K   Click here for Help.
L   T4:1's Preset Value.
M   T4:1's Accumulated Value.
N   Displays each timer element's Time Base.

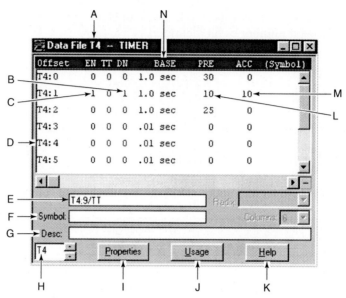

Figure 16-6 Timer Data File major features.

ACCESSING THE TIMER INSTRUCTIONS

As with other SLC 500 programming instructions, there are multiple ways to program an instruction on your ladder. Figure 16-7 illustrates the tabbed Instruction Toolbar with the Timer and Counter instructions tab selected. Note the instructions are identified by their mnemonics.

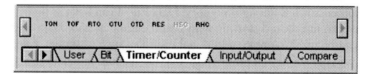

Figure 16-7 SLC 500 tabbed Instruction Toolbar showing the Timer/Counter tab contents.

The Instruction Palette is another option for selecting instructions and placing them on the ladder rung. As with the tabbed Instruction Bar, each instruction is identified by its mnemonic. Figure 16-8 illustrates the Instruction Palette identifying the Timer and Counter instructions as well as the tabbed Instruction Toolbar. Can you find the three timer instructions?

Callouts from Figure 16-8 are explained as follows:

A   TON Instruction.
B   Timer/Counter tab. Click here to display instructions.
C   Click here to display the Instruction Pallette.

PROGRAMMING THE TON TIMER

This exercise will guide you through developing a TON Timer ladder program, including the status bits. When finished developing the ladder logic, you will download the project to your SLC 500 PLC and run the application. As the application runs, you will monitor the ladder to understand how the counter and its associated status bits operate and interact. Figure 16-9 illustrates the ladder you will be programming for this exercise.

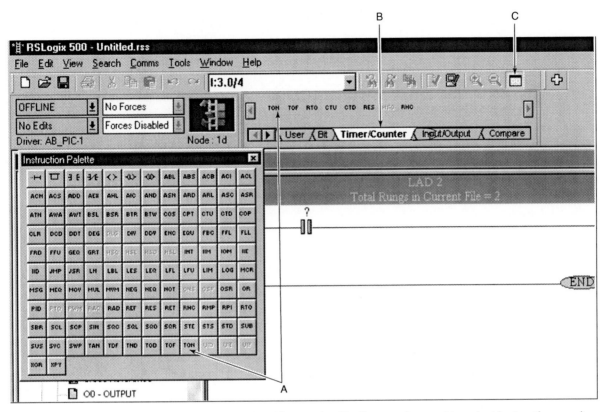

Figure 16-8   The Instruction Palette or the tabbed Instruction Toolbar can be used to select instructions and program them on a ladder rung.

THE LAB

1. _____ Open the Begin project created earlier.
2. _____ From the User tab, click on the Add New rung button.
3. _____ Add an XIC instruction and type in its address.
4. _____ Click on the Timer/Counter tab or Instruction Palette to view the instructions.
5. _____ Click on the TON instruction. The instruction should appear on the rung.
6. _____ Type in address T4:0.
7. _____ Press Enter.
8. _____ Select the 1.0 second time base out of the drop down box.
9. _____ Press Enter.
10. _____ Type in a Preset value of 30.
11. _____ Press Enter.
12. _____ Leave the Accumulated value as zero.
13. _____ Press Enter.
14. _____ To add the next rung, click on the User tab, or make a selection from the Instruction Palette.
15. _____ Click on the New rung button.
16. _____ Click on the XIC instruction.
17. _____ Type in the address T4:0/TT.
18. _____ Press Enter.
19. _____ Click on the OTE instruction.
20. _____ Type in address O:2/3.
21. _____ Press Enter.

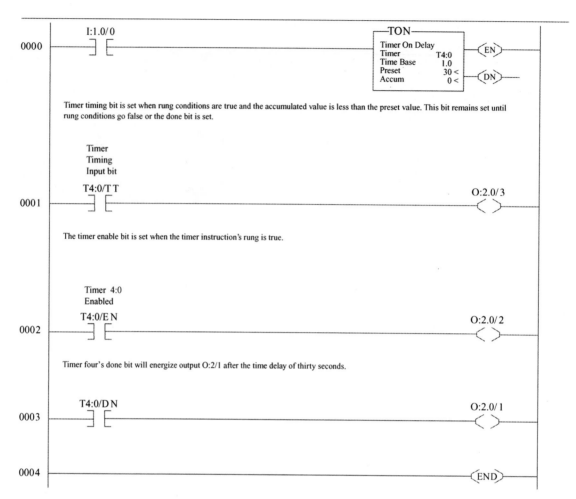

Figure 16-9 TON timer with associated status bits.

22. _____ Add rung comments as illustrated in Figure 16-9.
23. _____ To add the next rung, click on the User tab.
24. _____ Click on the New rung button.
25. _____ Click on the XIC instruction.
26. _____ Type in the address T4:0/EN.
27. _____ Press Enter.
28. _____ Click on the OTE instruction.
29. _____ Type in address O:2/2.
30. _____ Press Enter.
31. _____ Add rung comments as illustrated in Figure 16-9.
32. _____ Add the next rung.
33. _____ Enter rung comments.

Testing Your Project

With your timer program completed, download the program into your SLC processor, put processor in run mode, and go on-line. Energize input switch I:1/0. As your timer proceeds through the timing cycle, answer the following questions:

1. As long as input I:1/0 is true, the timer is _____.

2. Anytime a TON timer's input conditions go false, the timer _____.

3. The TT status bit is only true when _____.

4. Explain when the Done bit is true. What causes it to go false? _____

_____

_____

5. The TON timer is an on-delay timer. Explain what this means. What status bit is used and how is it programmed to use the TON instruction as an on-delay timer? _____

_____

_____

_____

6. Run your timer through its timing cycle again. As the timer begins, note the timer timing bit. Notice how the TT bit operates in relation to the DN bit. Do you see how you could use the TT and DN bits for different timing scenarios? Explain each. _____

_____

_____

_____

_____

7. Will the timer continue to time past its preset value? _____

8. If the timer instruction goes false anytime during its timing cycle, the timer will

_____.

9. _____ In the project tree, open the Data Files folder if it is not already open.

10. _____ Double-click on the T4 Timer file to open it. The Timer Data File, which is similar to Figure 16-6, should open.

11. _____ Run the timer again.

12. _____ Observe the Timer Status bits and Accumulated Value as the timer proceeds through its cycle.

PROGRAMMING THE SLC 500 TOF INSTRUCTION

Next we will program a TOF, or off-delay timer. This timer will allow us to run a fan for a specified time after a motor shuts off. A possible application could be a variable-frequency drive controlling a motor. One reason for using a drive is to enable the operator to run the motor at variable speeds. A typical induction motor runs at 1750 RPM when controlled by an across-the-line starter. At 1750 RPM the internal fan also travels at 1750 RPM and keeps the motor cool. When the same motor is placed on a variable-frequency drive, the motor may be operated at slow speeds for some applications. As a result, the cooling fan will also run at the same slow speed and not adequately cool the motor. In a situation such as this, an auxiliary fan would be used to cool the motor. We would want the auxiliary fan to come on at the same time as the motor starts, but we might wish to have the fan run for an additional time after the motor is turned off. The time the fan runs after the motor is turned off is called the off-delay. The TOF timer would be used to control the auxiliary fan. As you work through this section, keep this application in mind as you run your TOF timer project.

1. _____ Modify the program from Figure 16-9 to become the TOF program as illustrated in Figure 16-10.

2. _____ When the program is created, download, put processor in run mode, and go on-line.

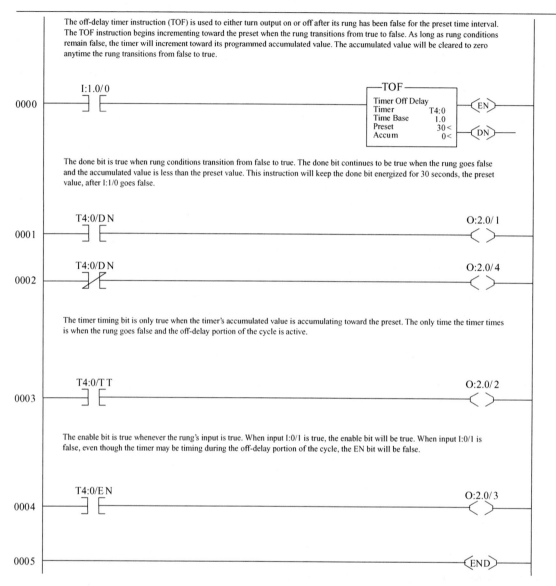

The off-delay timer instruction (TOF) is used to either turn output on or off after its rung has been false for the preset time interval. The TOF instruction begins incrementing toward the preset when the rung transitions from true to false. As long as rung conditions remain false, the timer will increment toward its programmed accumulated value. The accumulated value will be cleared to zero anytime the rung transitions from false to true.

The done bit is true when rung conditions transition from false to true. The done bit continues to be true when the rung goes false and the accumulated value is less than the preset value. This instruction will keep the done bit energized for 30 seconds, the preset value, after I:1/0 goes false.

The timer timing bit is only true when the timer's accumulated value is accumulating toward the preset. The only time the timer times is when the rung goes false and the off-delay portion of the cycle is active.

The enable bit is true whenever the rung's input is true. When input I:0/1 is true, the enable bit will be true. When input I:0/1 is false, even though the timer may be timing during the off-delay portion of the cycle, the EN bit will be false.

Figure 16-10 TOF ladder with status bits.

3. _____ As you energize the timer input, explain what happens. _____

_____

_____

_____

4. _____ Switch the timer's input to off or false. Does the timer start timing? _____

5. _____ Explain what you observe when the time expires. _____

_____

_____

_____

6. _____ When is the Enable bit true? _____

_____

_____

_____

7. _____ When is the TT bit true? _____

_____

_____

_____

8. _____ When is the Done bit true? _____

_____

_____

_____

9. _____ The TON timer resets itself whenever the input conditions controlling the timer
go false. What happens to the TOF timer when its input conditions go false?

_____

_____

_____

_____

PROGRAMMING A RETENTIVE TIMER (RTO)

In this lab exercise you will develop a timer that will work a lot like the TON, except this timer
will remember its accumulated time when its input conditions go false. This timer retains its status, thus it is called a retentive timer. Keep in mind that a PLC's retentive instructions only remember their status through a power interruption if their memory backup battery is good.

Since this timer is retentive it needs a separate Reset instruction (RES) to reset the timer
back to zero. You will be programming a reset instruction on rung 4 of the ladder program. You
will find the RES instruction either on the Timer/Counter tab on the tabbed Instruction Toolbar,
or in the Instruction Pallette. Remember to enter the address of the instruction that you wish to
reset. In this case you will be resetting T4:0. The Reset instruction will be used to reset instructions in addition to the RTO timer. We will work with the RES instruction when we work with
counters in the next section of this lab.

1. _____ Modify the program from Figure 16-10 to become the ladder rungs as illustrated in
Figure 16-11.
2. _____ Add complete documentation including a page title, assign symbols, and rung comments, explaining how each rung operates. Add address descriptions.
3. _____ When your project is completed, download the project to your processor. Put the
processor in Run mode and go on-line.
4. _____ As you test your program, answer the questions below:
A. When you energize input I:1.0/0 and then de-energize it five seconds later,

explain what you observe. _____

_____

_____

_____

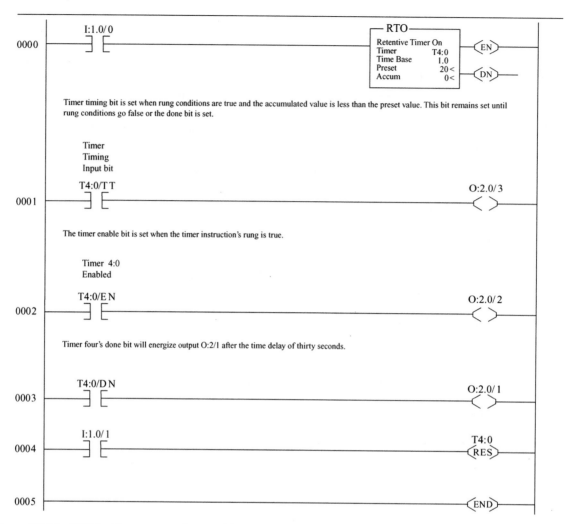

Figure 16-11 RTO ladder rungs and status bits.

B. Explain how the TT bit works. _____

_____

_____

_____

C. Explain if the DN bit operates any differently from when it is used with the TON

timer. _____

_____

_____

_____

D. How is the RTO timer reset? _____

_____

_____

_____

E. This timer retains its accumulated value and status bits through a shut down or power interruption. What makes this possible? _____

_____

_____

_____

5. _____ Check with your instructor as to whether your completed project is to be printed and handed in for credit.

TIMER PROGRAMMING EXERCISES

This section will provide you with hands-on experience developing and understanding additional timer applications.

Programming Exercise One

1. _____ Create the ladder rungs as illustrated in Figure 16-12. As you run this program, explain how this timer operates differently from the first TON timer you programmed earlier.

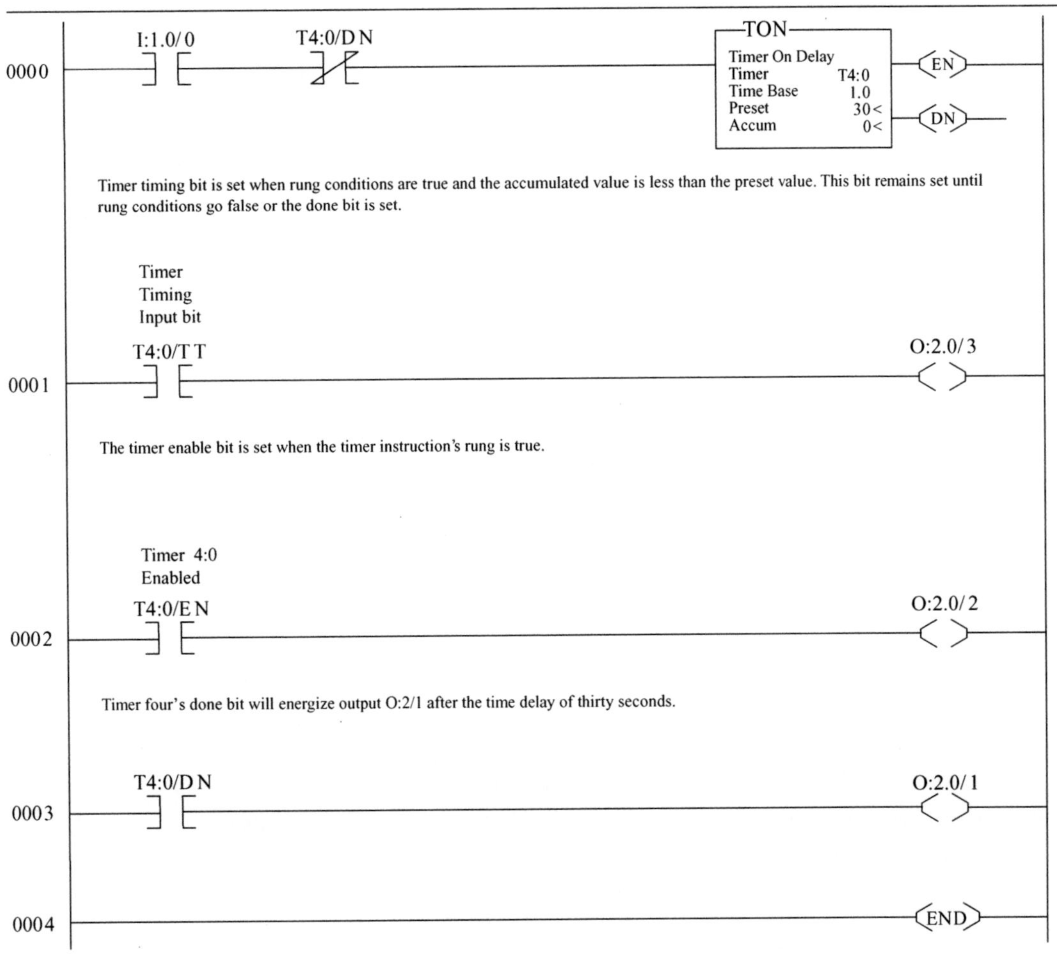

Figure 16-12 Timer program for Exercise One.

2. _____ Include completed documentation on your ladder.

3. _____ Question: Why does this timer behave the way it does? _____

_____

_____

_____

_____

4. _____ Check with your instructor as to whether your completed project is to be printed and handed in for credit.

Programming Exercise Two

Earlier we introduced the concept of using a timer and a variable-frequency drive auxiliary cooling fan. Since the motor was to run at slow speeds, we needed to have the auxiliary cooling fan continue to run after the motor was stopped.

1. _____ Develop the ladder logic so a common start-stop push-button station will control a variable-frequency drive. Since the motor may operate at slow speeds during the manufacturing process, incorporate a timer in your ladder logic so the auxiliary fan will continue to run five minutes after the motor stop push button is pressed.

2. _____ Add complete documentation explaining each rung and instruction. Include symbols for each instruction.

3. _____ When completed, download and run your project to verify proper operation.

4. _____ Check with your instructor as to whether your completed project is to be printed and handed in for credit.

5. _____ To save your timer project, use the Save As feature of Windows. Do not save the Begin project; discard changes.

Lab B: Programming Counters

OBJECTIVES

Upon completion of this laboratory exercise, you should be able to:

- identify the CTU and CTD counter instructions
- program a CTU counter and understand its associated status bits
- program a CTD counter and understand the operation of its status bits
- program the Reset instruction to reset the appropriate counter
- download the counter projects and go on-line with them
- create a counter application from a functional specification

INTRODUCTION

The SLC 500 family of PLCs has two counter instructions, Count-Up and Count-Down. Simply, the count-up counter would be used if you wanted to count up from 0 to 2,500. The count-down counter would be used to count down from 2,500 to 0. Since counters are retentive, a Reset instruction is used to reset the counter's accumulated value and associated status bits.

SLC 500 and MicroLogix 1000 counters are stored in counter file five in the data file section of the PLC's memory. The default counter file is file five. The counter file is used to store only counter data. Similar to timers, each counter consists of three 16-bit words, called a counter element. There can be many counter files for a single modular processor data file. (The MicroLogix 1000 only allows thirty-two counters in file five. No additional counter files or timer elements can be created.) Any data file greater than file eight can be assigned as an additional counter file. Each counter file can have up to 255 counter elements.

INSTRUCTOR DEMONSTRATION

Follow along as your instructor demonstrates an RSLogix 500 program containing the CTU and the CTD counters and their associated status bits.

TYPES OF COUNTER INSTRUCTIONS

The SLC 500 PLC family has the following counter instructions and a reset instruction. The table in Figure 16-13 identifies each counter instruction.

Figure 16-14 illustrates the Count-Up Counter, its associated status bits, and a Reset instruction.

Figure 16-15 illustrates the Count-Down Counter and its associated status bits.

COUNTER INSTRUCTIONS		
Instruction	**Use This Instruction to**	**Functional Description**
Count Up	Count from zero up to a desired value	Counting the number of parts produced during a specific work shift or batch. Also counting the number of rejects from a batch.
Count Down	Count down from a desired value to zero	An operator interface display shows the operator the number of parts remaining to be made for a lot of 100 parts ordered.
High-Speed Counter	Count input pulses that are too fast separately from normal input points and modules	Most fixed PLCs will have a high-speed set of input points that will allow interface to high-speed inputs. Signals from an incremental encoder would be a typical high-speed input. Check your specific PLC for the maximum pulse rate.
Counter Reset	Reset a timer or counter	Used to reset a counter to zero so another counting sequence can begin.

Figure 16-13   SLC 500 counter instructions.

UNDERSTANDING THE COUNTER INSTRUCTION

The counter instruction is comprised of several components. Refer to Figures 16-14 and 16-15 and identify the components of the counters.

- Mnemonic
- Identifies counter type
- The Counter parameter identifies the address in the counter file where the counter and its associated status information is stored.

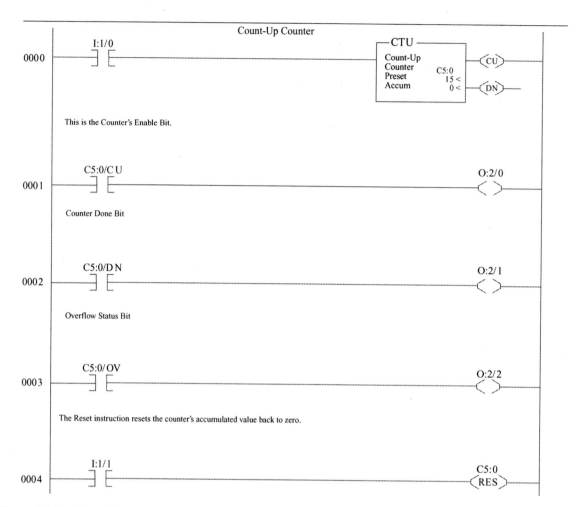

Figure 16-14 RSLogix count-up counter rung.

- The Preset is an integer value between −32,768 and +32,767 that identifies how far the counter is to count before it is considered done counting.
- The Accumulated value indicates our position in the counting cycle. The Accumulated value contains the current count stored in the counter.
- The CU is an indicator that displays whether or not the counter instruction is true, or enabled. The CU is one of the counter's status bits. Status bits will be explained in the next section.
- DN is also an indicator displaying when the counter is done counting. The DN is one of the counter's status bits.

COUNTER STATUS BITS

In order to monitor the operating status of the counter as it progresses through its counting cycle, status bits are available. Status bits can be programmed on ladder rungs and monitored, if desired, for specific conditions as identified below. Figure 16-16 describes the status bits for the CTU, or up-counter.

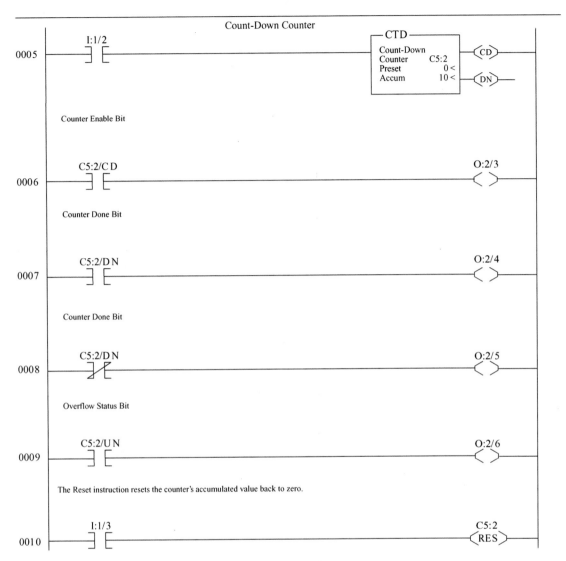

Figure 16-15 RSLogix count-down counter rung.

CTU COUNTER STATUS BITS			
Status Bit	**Identified as**	**Addressing Example**	**Functional Description**
Count-up enable	CU	C5:0/CU	This bit identifies if the count-up instruction is true or enabled.
Done	DN	C5:0/DN	When the accumulated value programmed for the instruction is equal to or greater than the programmed preset value, the Done bit will be true.
Overflow	OV	C5:0/OV	The counter has overflowed when the counter has exceeded the maximum value for a 16-bit signed integer, which is +32,767. If the accumulated value exceeds +32,767, the overflow bit is true. The counter wraps around and begins counting at a −32,768. The OV bit is a status bit only. The SLC 500 processor will not fault on a counter overflow.

Figure 16-16 CTU counter status bits.

Figure 16-17 describes the status bits for the CTD, or down-counter.

CTD COUNTER STATUS BITS			
Status Bit	**Identified as**	**Addressing Example**	**Functional Description**
Count-down enable	CD	C5:0/CD	This bit identifies if the count-down instruction is true or enabled.
Done	DN	C5:0/DN	When the accumulated value programmed for the instruction is greater than or equal to the programmed preset value, the Done bit will be true.
Underflow	UN	C5:0/UN	The counter has underflowed when the counter has exceeded the maximum negative value for a 16-bit signed integer, which is $-32,768$. If the accumulated value exceeds $-32,768$, the underflow bit is true. The counter wraps around and begins counting at $+32,767$. The UN bit is a status bit only. The SLC 500 processor will not fault on a counter underflow.

Figure 16-17   CTD counter status bits.

SLC 500 COUNTER DATA FORMAT AND ADDRESSING

Counters are three-word elements. Word zero contains counter status bits. Word one contains the preset value, and word two contains the accumulated value. Counter instruction data is in one of two data formats, bits or words. Status bit data will be single-bit data, while preset and accumulated value data will be represented as a 16-bit signed integer. Figure 16-18 illustrates the format of the counter element.

Word 0	CU	CD	DN	OV	UN	UA	Reserved Bits	
Word 1	Preset Value							
Word 2	Accumulated Value							

Figure 16-18   RSLogix 500 counter element three-word format.

Figure 16-18 illustrates a counter instruction or element. Each counter element has three separate words, one containing status bits and two (a 16-bit signed integer containing word data) representing the preset and accumulated value. These two words are subelements of the counter element. A subelement is simply a separate piece of an element or instruction that is addressable as a stand-alone unit. The preset value and accumulated value can be addressed individually for use in your ladder program. As an example, when producing product A, a case consists of 24 pieces. In this situation, the counter tracking the number of pieces packed in each case will have a preset value of 24. The next order is for product B, which has 12 pieces per case. The operator sets the machine to produce product B, and a ladder rung using a Move instruction will move the new value of 12 into the preset value subelement of the counter instruction.

The addressing format for counters is very similar to timer address format. Examples of counter instruction status bit and subelement components are listed in Figure 16-19.

COUNTER ADDRESS FORMAT	
Counter Address	**Address Identifies**
C5:0/DN	Counter file 5, element 0, the done bit
C5:46/CU	Counter file 5, element 46, count-up enable bit
C12:9/OV	Counter file 12, element 9's overflow bit
C21/15.PRE	Counter file 21, element 15's preset value
C5:1.ACC	Counter file 5, element 1, the accumulated value

Figure 16-19   RSLogix 500 counter addressing format.

ACCESSING THE COUNTER INSTRUCTIONS

As with other SLC 500 programming instructions, there are multiple ways to program a counter instruction on your ladder rung. Figure 16-20 shows the Timer/Counter tab (A) from the tabbed Instruction Toolbar. The CTU instruction (B) as well as the CTD (C) and Reset instruction (D) are identified.

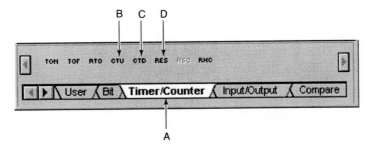

Figure 16-20   RSLogix 500 tabbed Instruction Toolbar showing the Timer/Counter tab.

Figure 16-21 shows the Instruction Pallette and identifies the CTU instruction (A), CTD instruction (B), and RES instruction (C).

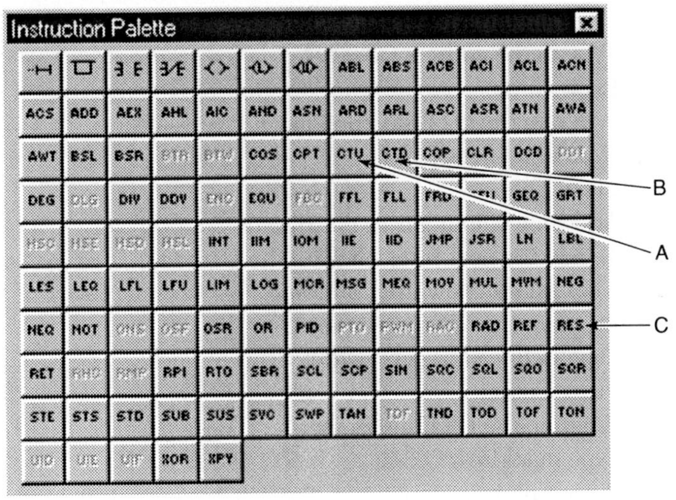

Figure 16-21   RSLogix 500 Instruction Pallette identifying the CTU, CTD, and RES instructions.

PROGRAMMING THE CTU COUNTER

This exercise will guide you through developing an up-counter ladder program, including the status bits. When finished developing the ladder logic, you will download the project to your SLC 500 PLC and run the application. As the application runs, monitor the ladder to understand how the counter and its associated status bits operate and interact. Figure 16-14 illustrates the ladder you will be programming for this exercise.

THE LAB

1. _____ Open the Begin project created earlier.
2. _____ From the User tab, click on the Add New rung button.
3. _____ Add an XIC instruction and type in its address.
4. _____ Click on the Timer/Counter tab to view the instructions.

5. _____ Click on the CTU instruction. The instruction should appear on the rung.
6. _____ Type in address C5:0.
7. _____ Press Enter.
8. _____ Type in a Preset value of 15.
9. _____ Press Enter.
10. _____ Leave the Accumulated value as zero.
11. _____ Press Enter.
12. _____ Add rung comments as illustrated in Figure 16-14.
13. _____ To add the next rung, click on the User tab.
14. _____ Click on the New rung button.
15. _____ Hold down the left mouse button and drag the XIC instruction into position.
16. _____ Type in the address C5:0/CU.
17. _____ Press Enter.
18. _____ Hold down the left mouse button and drag the OTE instruction into position.
19. _____ Type in address O:2/0.
20. _____ Press Enter.
21. _____ Add rung comments.
22. _____ To add the next rung, click on the User tab.
23. _____ Add the new rung.
24. _____ Program the XIC instruction.
25. _____ Type in the address C5:0/DN.
26. _____ Press Enter.
27. _____ Program the OTE instruction.
28. _____ Type in address O:2/1.
29. _____ Press Enter.
30. _____ Add rung comments as illustrated.
31. _____ To add the next rung, click on the User tab.
32. _____ Click on the New rung button.
33. _____ Program the XIC instruction.
34. _____ Type in the address C5:0/OV.
35. _____ Press Enter.
36. _____ Program the OTE instruction.
37. _____ Type in address O:2/2.
38. _____ Press Enter.
39. _____ Add rung comments as illustrated.

Resetting the CTU Counter

Since counters are retentive, the Reset instruction is needed to reset the counter's Accumulated value back to zero. The Reset instruction used to reset our counter is the same RES instruction used to reset the RTO instruction.

40. _____ Program the Reset instruction as illustrated.
41. _____ Energize the input controlling the RES instruction. The counter Accumulated value should reset to zero.
42. _____ Energizing the input controlling the RES Instruction will reset the counter accumulated value to zero.

Testing Your Project

With your counter program completed, download the program into your SLC processor, put the processor in run mode, and go on-line. Increment input switch I:1/0. As your counter increments, answer the following questions:

1. As the counter increments, explain the operation of the CU bit. _____

_____

_____

_____

2. When the preset value is _____, the accumulated value of the Done bit is true.

3. The data range for the counter's preset value or accumulated value is _____.

MONITORING THE COUNTER DATA FILE

1. _____ Open the counter data file.
2. _____ As you increment your counter, monitor the file. The major sections of the file are called out in Figure 16-22.

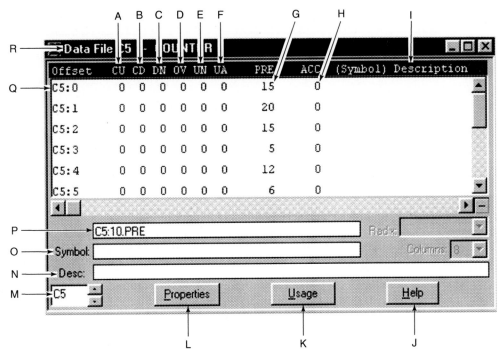

Figure 16-22 RSLogix 500 Counter Data File C5.

A	Count-Up enable status bit
B	Count-Down enable status bit
C	Done status bit
D	Overflow status bit
E	Underflow status bit
F	High-speed counter status bit (MicroLogix only)
G	Preset value
H	Accumulated value
I	If a Symbol has been entered, it will display here.
J	Click here for help.
K	Click here to view Counter Usage.
L	Click here to view and modify Counter Properties.

M   Displays current data file being viewed. Click on up or down arrows to move to another data file.

N   Add a Text Description to the currently selected counter.

O   Add Symbol here.

P   Identifies the cursor as on C5:10.PRE (not shown in window view).

Q   Counter element C5:0.

R   Identifies data file being viewed.

As you monitor the counter data file, answer the questions below.

1. Which counter status bits are valid for the current counter application? _____
_____

2. What is the data range of the Preset and Accumulated values? _____

3. What will be found on the Properties window? _____
_____
_____

4. Open the Usage window and explain what information this window provides. Explain the terminology displayed in the window when an instruction is or is not used. _____
_____
_____
_____

HINT: Consult the Help screens from the Usage window.

OVERFLOWING THE CTU COUNTER

In this section, we will overflow the counter and learn how the processor behaves as the counter overflows.

1. _____ With your processor still running, open the counter's data file, C5. The opened file is illustrated in Figure 16-23.

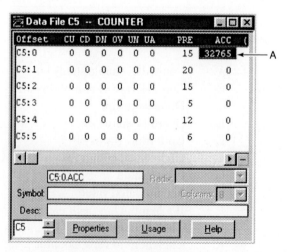

Figure 16-23 Adjusted preset value for counter C5:0.

2. _____ Double-click on the ACC value for C5:0.

3. _____ Type in the value 32765 into the Accumulated value (A) of C5:0, as illustrated in Figure 16-23.

4. _____ Press Enter.

5. _____ Close the Counter data file window.

6. _____ Continue to increment the counter until it counts one more than 32767.

7. _____ Explain what you observe.

    A. What is the next counter accumulated value after 32767? _____

    B. Explain the current status of your status bits.

       CU: _____

       DN: _____

       OV: _____

    C. Is the processor faulted? _____

8. _____ If you continue to increment the counter, explain how the Accumulated values

count. _____

_____

_____

_____

PROGRAMMING THE SLC 500 CTD INSTRUCTION

Next we will program a CTD counter. This counter will enable us to count, as an example, from 100 down to 0. For this exercise, assume that the current counter value will be output to a data display so the operator will know how many pieces are left to produce to complete this order.

1. _____ Enter the CTD program as illustrated in Figure 16-15.

   HINT: You might wish to program the ACC value at 10 so you will not have to decrement the counter 100 times to reach 0.

2. _____ When the program is created, download, put processor in run mode, and go on-line.

3. _____ As you decrement the counter through 0, explain the behavior of the DN bits.

4. _____ Energize the RES instruction.

5. _____ Explain what you observe. _____

_____

_____

_____

6. _____ Is it correct to reset your CTD counter so you can decrement as the next 10-piece

order is manufactured? _____

7. _____ If not, how do you suggest we reset our counter so we can start the next order?

_____

_____

_____

Since the Reset instruction resets the accumulated value to 0, the Reset instruction will not be of much value in this application. In order to start another order, the value 10 needs to be replaced in the counter Accumulated value. The address of the Accumulated value is C5:0.ACC. The Move instruction is a data manipulation instruction that will be used to move the value 10 into C5:0.ACC. The Move instruction source value will be the constant 10. When true, the instruction will place (move) a copy of the source value into the destination, C5:2.ACC. We will work with the Move instruction more in the next lab exercise.

8. _____ Go off-line and edit the project.
9. _____ Edit the rung currently containing the Reset instruction. Delete the Reset instruction. Refer to Figure 16-24.

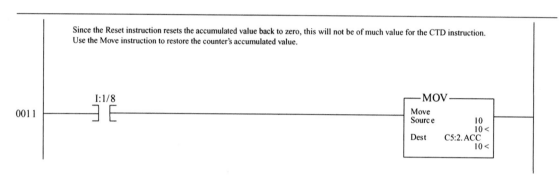

Figure 16-24 Move instruction will replace the value 0 in C5:2.ACC with the starting value of 10.

10. _____ Go to the Move/Logical tab on the tabbed Instruction Toolbar, or display the Instruction Palette.
11. _____ Select the MOV instruction.
12. _____ Program the Move instruction in place of the Reset instruction.
13. _____ Click on the Move instruction source parameter.
14. _____ Type in the value 10.
15. _____ Press Enter.
16. _____ Type in C5:2.ACC as the destination address for the value 10.
17. _____ Press Enter.
18. _____ When finished, download your project, put the processor in run mode, and go on-line.
19. _____ Reset your counter using the Move instruction. Does it work better now? _____

PROGRAMMING A COUNTER THAT COUNTS UP AND DOWN

In this lab exercise, you will develop a counter pair that will count up and down at the same time. The scenario is that you want a running total of the total good parts produced. The CTU counter counts the pieces as they are made. The CTD decrements whenever there is a part rejected from the production line.

1. _____ Enter the ladder rungs as illustrated in Figure 16-25.
2. _____ When your project is completed, download the project to your processor. Put the processor in run mode and go on-line.

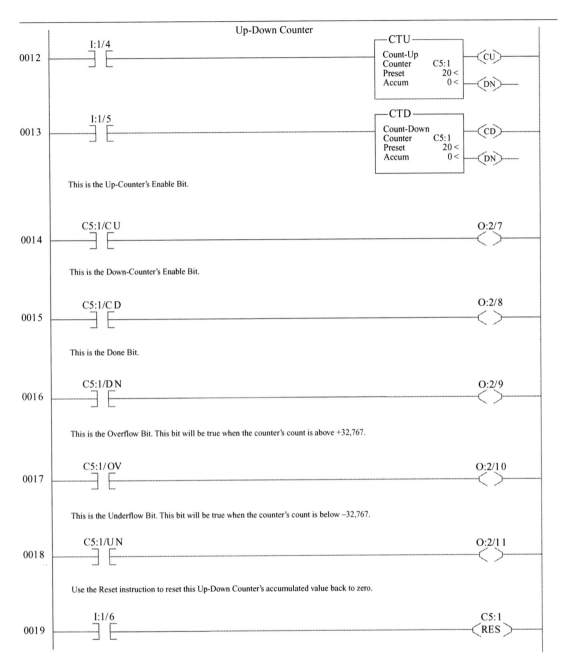

Figure 16-25   RSLogix 500 Up-Down Counter pair.

3. _____ As you test your program, refer to Figure 16-25 and answer the following questions:

   A. Explain why these two counters work together. _____

   _____

   _____

   _____

   _____

B. Why does the Reset instruction work in this application when it did not correctly reset the down-counter in an earlier lab exercise? _____

_____

_____

_____

COUNTER PROGRAMMING ASSIGNMENTS

Complete the following counter programming assignments referring to the functional specification for each.

Programming Exercise One

For this exercise, we have a manufacturing scenario where completed product is coming down the conveyor to be put in shipping cases. There are to be 24 products per case. A photo eye on the conveyor counts the pieces as they go into a case. When a case is full, it is to be moved down the line and a new case moved into position.

1. _____ How are you going to control the mechanical action of moving the full case down the

   line so another case can be positioned and filled? _____

   _____

   _____

   _____

   _____

2. _____ Reset the counter so another case of 24 pieces can be filled.
3. _____ Have the first counter trigger a second counter to count the number of cases produced.

Programming Exercise Two

You have been hired to develop a ladder project to control the operation of a parking garage. Each time a car enters the driver takes a ticket and proceeds to find an available parking spot. You need to develop a system to know when there are parking spaces available and when there are none. If there are spots available, the ticket machine issues a ticket and allows the car into the lot. Assume the parking lot holds 20 cars. If the lot is full, no ticket is issued, the arm across the entry is not raised, and the LOT FULL sign is illuminated.

We need to count each time a car enters the lot and when a car exits the lot. This can be accomplished by having a sensing device counting cars as they exit and enter the lot.

The following devices connect to our PLC:

- Ticket dispensing machine signal to dispense ticket
- Arm allowing or restricting entry to the lot
- LOT FULL sign, designed to turn ON or OFF
- Sensor looking for the presence of a car at the ticket dispenser
- Sensor watching for cars to exit the lot

Which devices would send input signals to our PLC, and which devices would receive output signals from our PLC? Fill in the table in Figure 16-26 indicating if the device is an input or output. Assign addresses and symbols you intend to use as you develop your ladder program.

PARKING GARAGE I/O ASSIGNMENT SHEET			
Device	Input or Output?	Address	Symbol
Ticket dispensing machine			
LOT FULL sign			
Sensor looking for the presence of a car			
Arm allowing or restricting entry to the lot			
Sensor watching for cars to exit the lot			

Figure 16-26   I/O assignments for parking garage ladder program.

Develop the parking garage application as a new project.

1. _____ Open up the RSLogix 500 software.
2. _____ Create a new project.
3. _____ Configure the PLC's I/O.
4. _____ Verify power supply loading.
5. _____ Verify that the processor's channel configuration is correct.
6. _____ Do not use the default counter file 5. Create a new counter file, and use counters from that file. If using a MicroLogix 1000, you will not be able to create a new counter file; use file 5.
7. _____ As you create your ladder logic, add page titles, rung comments, and symbols for each instruction and rung.
8. _____ When the ladder is complete, download the project to your PLC.
9. _____ Put the PLC in RUN Mode and go on-line.
10. _____ Interpret the PLC ladder rungs and answer questions 11 through 18.
11. _____ Explain the function of the Preset parameter for the counter instruction for this

   particular application. _____

   _____

   _____

12. _____ What does the Accumulated value tell us about the parking lot? _____

   _____

13. _____ Switch _____ simulates a car entering the parking lot. Press this switch ON and OFF to increment the counter. Does the Accumulated value increment once

   for each on and off cycle of the switch? _____

14. _____ Is a ticket dispensed (simulated by output _____), and the gate arm raised

   (simulated by output _____) each time a car enters the lot and the lot is not

   full? _____

15. _____ Increment the counter until the lot is full. Does the lot full sign output come on?

   _____

16. _____ Since the lot is full, if a car attempts to enter, is a ticket dispensed? _____

17. _____ Does the gate arm raise to allow the car to enter? _____

18. _____ What status input is used to make things happen when the lot is full? Why do we use

   this particular signal? _____

   _____

19. _____ When enough cars leave the parking lot so that new cars can enter, verify your program for proper operation as new cars attempt to enter.

20. _____ Check with your instructor to see if a check off is needed at the completion of this exercise.

21. _____ Check with your instructor to see if a printout of your project needs to be generated and handed in for credit.

REVIEW QUESTIONS

1. A counter is made of three words called a counter _____.

2. Explain when the DN bit is true. _____

_____

_____

3. Describe the difference between the CU and CD status bits. _____

_____

_____

4. What is the data range of a counter preset or accumulated value? _____

5. List the counter subelements. _____

_____

_____

6. If a counter overflows or underflows, will the processor fault? _____

7. List each of the counter status bits. Include a sample address and a detailed description of each bit.

8. List the two counter subelements. Include a sample address and definition of each. _____

_____

_____

9. What is the function of the reset instruction? _____

_____

10. Explain why a RES instruction is or is not used to reset the CTD. _____

_____

_____

11. Identify the parts of the counter file shown in Figure 16-27.

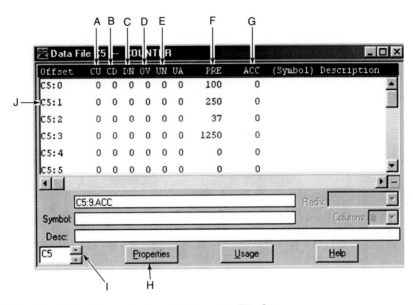

Figure 16-27   Identifying the parts found in Counter File C5.

A. _____

B. _____

C. _____

D. _____

E. _____

F. _____

G. _____

H. _____

I. _____

J. _____

12. Are counters retentive or non-retentive? What does this mean? _____

_____

_____

13. How many counter elements can be created in a single counter file? _____

14. How many total data files can be created in an SLC 500 modular processor, assuming adequate memory? _____

15. List the steps to create a user-defined counter file.

16. When looking at a counter data file, six status bits are identified for each element. Are all six status bits used with all counter instructions? Explain why or why not. _____

_____

_____

_____

17. What would happen if the MOV instruction shown in Figure 16-28 was executed and the counter was incremented? _____

_____

_____

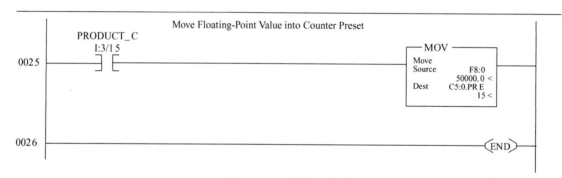

Figure 16-28 Moving the value in F8:0 to C5:0 PRE.

A. Assume F8:0 contains 50,000.
B. Ask your instructor if you should verify your answer by developing and testing the ladder rungs.

17

Comparison and Data Handling

Lab A: The Copy Instruction

OBJECTIVES

Upon completion of this laboratory exercise, you should be able to:

- understand the Copy File instruction
- understand how to use the RSLogix 500 File Usage feature

INTRODUCTION

The copy instruction copies blocks of data from one data file location to another data file location. This is called a file-to-file transfer. Data in the original location, called the source, is left intact while a copy of the source information is copied to the designated destination. Copying a group of words of like data is called copying a *user-defined file*. Data can be copied to another location either within the same data file, or to another data file. A user-defined file could contain recipe data of all the ingredients needed to make a batch of chocolate chip cookies. The copy instruction could be used to copy the chocolate chip recipe into the proper PLC data table so that chocolate chip cookies can be made when the current batch of peanut butter cookies is completed. The copy of the peanut butter cookie recipe stored in the working PLC data file will be written over when the copy instruction copies the chocolate chip cookie recipe to the working data file recipe.

COPY INSTRUCTION PARAMETERS

The copy instruction has three parameters: source, destination, and length. *Source* is the address of the file to be copied; *destination* is the starting address of the data file the information is to be copied to; *length* is the number of elements you want to copy. The destination file type determines the number of words the instruction will copy. Refer to the text for additional information on the copy instruction and data types and how they affect the number of words or elements copied.

COPY INSTRUCTION PARAMETER SYNTAX

The *Copy* instruction is used to copy a block of data from one data file location to another. A user may group like data together in consecutive memory locations within a data file. As an example, a group of integer file locations containing recipe data for timing, counting, weight, mixing times, etc., would be stored together as like information pertaining to one recipe. There may be many blocks of data, each a single recipe, stored in a data file such as the integer file. Each block of data within such a file is called a user-created file.

When you want to copy a recipe with, as an example, twelve recipe values, not only do the source and destination need to be specified, the source and destination must be identified as a user-created file so the CPU will know that it is to copy a consecutive block of data. To identify a block of data as a user-created file, the # sign is placed before the source and the destination parameters. An example of a copy instruction is illustrated in Figure 17-1.

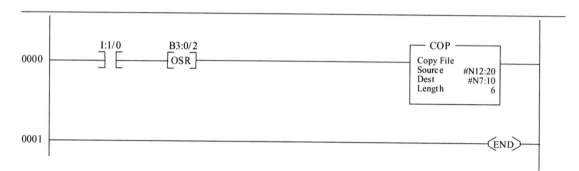

Figure 17-1 The RSLogix 500 Copy File instruction.

The *Copy* instruction illustrated specifies a source user-created block of data, called a user file, starting at N12:20 (integer file twelve, element twenty) to be copied to the starting file address N7:10 (integer file seven, element ten). The length of the file is six elements. Since integer file elements are one word in length, six elements will be copied from the source to the destination. Remember, the destination data file type determines the number of elements/words that will be copied.

COPY INSTRUCTION STATUS BITS

Copy instruction execution does not affect any status bits.

THE LAB

This lab exercise will develop the ladder program to copy the recipe data from integer file twelve, word twenty, to integer file seven, word ten. The recipe contains six recipe parameters for time, count, weight, and mixing times. For this application, assume recipe values for integer file N12:20 are as follows: 10, 20, 30, 25, 33, 12. See Figure 17-2.

#N12:20				#N7:10	
N12:20	10	→	10	N7:10
N12:21	20	→	20	N7:11
N12:22	30	→	30	N7:12
N12:23	25	→	25	N7:13
N12:24	33	→	33	N7:14
N12:25	12	→	12	N7:15

Figure 17-2 Conceptual illustration of how the Copy File instruction will operate on data.

1. _____ Open the Begin project.
2. _____ Create a new Integer file, N12.
3. _____ Assign enough elements to the new file as needed for this lab exercise.
4. _____ Enter integer data into N12 as illustrated in Figure 17-2. The completed N12 file should look similar to Figure 17-3.

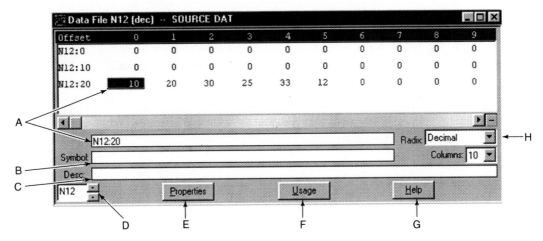

Figure 17-3 N12 integer file with source data entered.

Figure 17-3 features are listed as follows:

A	Address N12:20
B	Enter a Symbol here for the listed file element if desired.
C	Enter a Description here for the listed file element if desired.
D	Use up and down arrows to move between data files.
E	Click to view file Properties.
F	Click to view file element Usage.
G	Click for Help.
H	Change file Radix.

5. _____ Click on the Usage button to view file usage. After the file element(s) are used in a ladder file instruction, an X will be placed in the element address, identifying the element as being used. This is a useful feature when programming and attempting to determine available elements. Figure 17-4 illustrates the N12 file after clicking the usage button (A). Notice the X in addresses N12:20 (B) through N12:25. When finished viewing the usage, click Data File (C) to return to the Data File View (top).
6. _____ Open Integer file, N7. Expand the file so there are enough elements for this lab exercise.
7. _____ Program the Copy File instruction on your ladder.
8. _____ Verify your project when finished programming.
9. _____ Open file N12.
10. _____ Click on Usage.
11. _____ Close the data file.
12. _____ Download, run the program, and go on-line.

To monitor the data files as Copy File instruction is executed:

13. _____ From the Windows menu bar, click on View.
14. _____ Close the standard, on-line, and tabbed Instruction Toolbar. This will close these toolbars, thus leaving additional room on your computer screen to view two data files, N12 and N7.

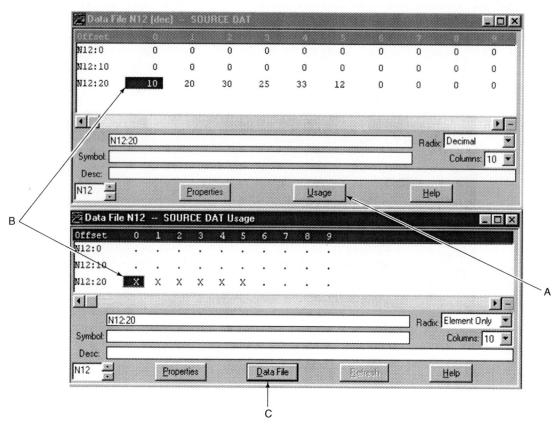

Figure 17-4   Displaying the data file and the usage of the N12 integer file.

15. _____ Before testing the project, open the N12 file.
16. _____ Move the data file to the top right of your screen.
17. _____ Open file N7. Move the file window below the N12 file. Resizing will be necessary.
18. _____ As you execute the Copy instruction, watch the data file displays on your screen.
19. _____ Fill in the table in Figure 17-5 with the data copied to file N7.
20. _____ Save project as Copy 1.

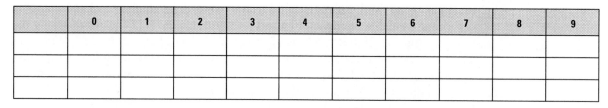

	0	1	2	3	4	5	6	7	8	9

Figure 17-5   Integer file N7 data as a result of execution of the Copy instruction.

REVIEW QUESTIONS

1. Explain why the OSR instruction is part of the ladder diagram. _____

_____

2. What would happen if I:1/0 were left energized and no OSR instruction was programmed on
the ladder rung? _____

3. What keystrokes did you use to monitor data file seven? _____

4. While monitoring N7, what function key would you press to go directly to N12? _____

5. Explain the significance of the # character in the copy file instruction. _____

_____

6. What are the three copy instruction parameters? _____

7. Define the first copy instruction parameter. _____

_____

8. Define the second copy instruction parameter. _____

_____

9. Define the third copy instruction parameter. _____

_____

10. Explain the role of status bits in conjunction with the copy instruction. _____

_____

Lab B: Copy File Programming

OBJECTIVES

Upon completion of this laboratory exercise, you should be able to:

- use the Copy File instruction to practice copy recipes
- create and expand new data files to hold recipe data
- enter recipe data into an integer file

INTRODUCTION

One popular use of the Copy instruction is for setting up recipe data for different products just before they are manufactured. Let's assume we manufacture four different types of soft drinks. Our four drinks are fruit punch, tropical punch, citrus punch, and orange. Each product has a different recipe. Each recipe contains data regarding the amount of each ingredient that is needed to produce the product. Figure 17-6 lists the ingredients, in gallons, in each drink recipe.

Each list of ingredients for each drink product is its recipe. Since each amount is an integer value, we will store these values in integer file N12. We have arbitrarily selected and will need to create integer file N12 for two reasons. First, we want to keep all the recipes together. Second, we want to keep the user-defined recipe files separate from the standard or working integer file, N7. This is done only to help avoid confusion and accidents of data being written over when mixing single integer storage with multiple integer (recipe) storage. We will reserve ten words to store data for each recipe.

FRUIT DRINK RECIPES				
Ingredients	**Fruit Punch**	**Tropical Punch**	**Citrus Punch**	**Orange**
Water	100	120	130	150
Sweetener	25	25	25	20
Grape Flavor	8	9	0	0
Orange Flavor	0	25	75	90
Pineapple Flavor	10	25	30	0
Apple Flavor	2	8	0	0
Pear Flavor	1	0	1	0
Strawberry Flavor	8	0	0	0
Passion Fruit Flavor	0	10	0	0

Figure 17-6   Each recipe contains data regarding the amounts of each ingredient in gallons needed to produce the product.

To keep this example simple, the operator will select which recipe will be manufactured by positioning a four-position selector switch. Figure 17-7 contains data regarding our input address allocation and data file allocation. Complete the table in Figure 17-7 with information pertaining to your specific PLC set up.

INPUT ADDRESS ASSIGNMENT WORKSHEET			
Drink	**Selector Switch Input Address**	**Recipe Storage Addresses**	**Total Words Reserved for Recipe**
Fruit Punch			10
Tropical Punch			10
Citrus Punch			10
Orange			10

Figure 17-7   Input Address assignment worksheet.

Figure 17-8 is an example of the selection portion of our drink manufacturing PLC ladder program. Notice that each selector switch position energizes a copy instruction. Each copy instruction copies the correct recipe to the working data files, N7:50 in our program.

EQUIPMENT REQUIRED

You will need a four-position selector switch connected to your PLC input section. Each switch position should have one circuit for each contact closure.

THE LAB

1. _____ Hardware Hookup
   Hook up a four-position selector switch to the input addresses listed in Figure 17-7, your Input Address Assignment Worksheet. If the selector switch is not available, connect four toggle switches to the four inputs to simulate the four-position selector switch.
2. _____ Open your Begin project.
3. _____ Create Integer file N12.
4. _____ Assign Integer file N7 a minimum of 75 elements.

NOTE: If using a MicroLogix 1000, your integer file 7 is the only integer file available. The MicroLogix 1000 integer file contains a fixed maximum of 105 elements. You cannot create integer file N12, or any additional data files. Use N7 and store the recipe starting at N7:0. Use N7:50 as the beginning destination address.

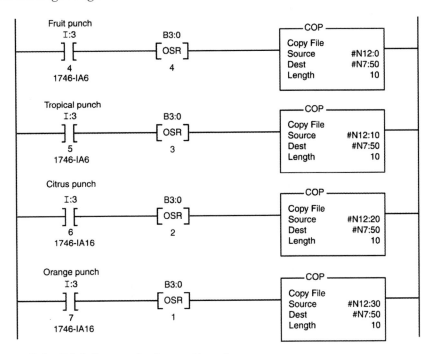

Figure 17-8   Drink flavor selection portion of our program.

5. _____ Enter recipe data into the source file as listed in Figure 17-6. Refer to Figure 17-20 in your textbook. Make sure you start with the water value as the first value for each recipe.

6. _____ At this point, the destination file should contain all zeros.

7. _____ Create program rungs as illustrated in Figure 17-8. Make certain your program addresses are correct for your PLC set up.

8. _____ Add symbols as illustrated in Figure 17-8.

9. _____ Add rung comments explaining each rung's function.

10. _____ When finished programming, download and save as recipe. Go on-line with the project and verify proper operation.

11. _____ Open the Destination Integer file and observe each recipe being copied as each input condition is selected.

12. _____ Ask your instructor if your exercise execution needs to be checked off.

13. _____ Ask your instructor if this exercise is to be printed and handed in.

REVIEW QUESTIONS

1. What kind of data can you store in N7? _____

2. Explain what the copy instruction does. _____

3. Illustrate how this file operates on the two integer files we will be using in this lab exercise.

_____

_____

4. Which status bits are associated with this instruction? _____

5. Explain how status bits can be used in conjunction with this instruction. _____

_____

6. Define the source parameter as it pertains to this instruction. _____

_____

7. Explain when the # symbol must be used in conjunction with this instruction. _____

_____

8. When the # is used, explain how it pertains to the parameters. _____

_____

9. How is the length parameter used with the copy file instruction? _____

10. What is the maximum number of words that can be copied in one block? _____

11. Compose a short explanation on how the instruction and file boundaries interact. _____

_____

_____

Lab C: Programming the Move Instruction

OBJECTIVES

Upon completion of this laboratory exercise, you should be able to:

- understand the workings of the Move instruction
- incorporate the Move instruction into an application to provide dynamic data to a counter preset value so the preset value can be changed "on the fly" as the product mix changes

INTRODUCTION

This exercise will introduce you to the *Move* instruction and how it is used to modify rung conditions as required by changing production conditions.

Each selector switch position will close a contact block which will be wired into a PLC input screw terminal as follows:

Wire selector switch position one contact block to PLC input I:1/0.

Wire selector switch position two contact block to PLC input I:1/1.

Wire selector switch position three contact block to PLC input I:1/2.

Each selector switch position will make one of three *Move* instructions true. Each *Move* instruction will use an integer value stored in an integer file location. When a specific *Move* instruction is selected, its associated integer value will be placed in a counter's preset value. The counter preset will change in relation to the selector switch position and the value in the associated *Move* instruction. The program is illustrated in Figure 17-9.

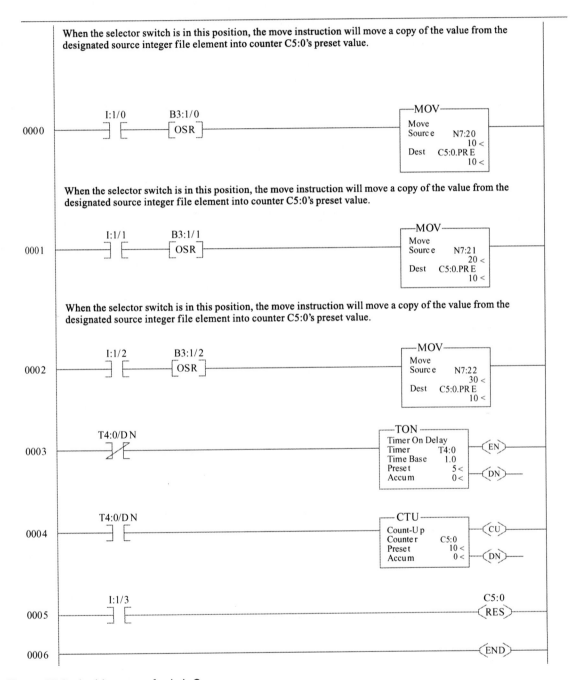

Figure 17-9   Ladder rungs for Lab C.

EQUIPMENT REQUIRED

A three-position selector switch connected to your PLC input section. Each switch position should have one circuit contact closure.

THE LAB

1. _____ Connect the three-position selector switch to your PLC input screw terminals as outlined in the introduction for this lab exercise.
2. _____ Open the Begin project.
3. _____ Create the necessary elements in Integer file 7.
4. _____ Create the ladder program as illustrated in Figure 17-9.
5. _____ Assign symbols to each input.
6. _____ Add rung comments as illustrated.
7. _____ Enter the Move instruction source data into the Integer file as listed below.

N7:20	10
N7:21	20
N7:22	30

8. _____ When finished programming, download and save as move. Put the processor in run mode and go on-line with the project.
9. _____ As the program runs, move the selector switch to different positions. Observe that the count preset values change as the commanded values are moved into them.
10. _____ Ask your instructor if your exercise execution needs to be checked off.
11. _____ Ask your instructor if this exercise is to be printed and handed in.

REVIEW QUESTIONS

1. Describe the operation of the Move instruction. _____
   _____
   _____
   _____

2. Define what types of data may be programmed into the source parameter. _____
   _____
   _____

3. Define what types of data may be programmed into the destination parameter. _____
   _____
   _____

4. What is the destination address C5:0.PRE? _____

5. Why is there a decimal point between the C5:0 portion of the address and the PRE?
   _____

6. Explain the operation of the OSR instruction on your ladder rungs. _____
   _____
   _____
   _____

7. Why would the OSR instruction be used in conjunction with the MOV instruction? _____
_____
_____

8. Describe the difference between the Move and Copy instructions. _____
_____
_____

9. List five situations where the move instruction might be used in a real manufacturing environment as programmed here.

10. Answer the questions below based on data in the sample integer data file in Figure 17-10.

Address	0	1	2	3	4	5	6	7	8	9
N7:0	1	2	3	4	5	6	7	8	9	10
N7:10	100	110	120	130	5	10	15	20	25	30
N7:20	10	20	30	0	0	4	6	9	1	4
N7:30	2	4	6	8	10	12	14	16	18	20
N7:40	20	25	30	35	40	45	52	63	74	89

Figure 17-10   Table data for question 10.

1. N7:10 = _____

2. N7:25 = _____

3. N7:42 = _____

4. N7:2 = _____

5. N7:9 = _____

6. N7:0 = _____

7. N7:33 = _____

8. N7:46 = _____

Lab D: Comparison Instructions

OBJECTIVES

Upon completion of this laboratory exercise, you should be able to:

- program a free-running timer and understand its operation
- understand operation of the comparison instructions

INTRODUCTION

This exercise will introduce you to incorporating comparison instructions to trigger events at counter accumulated values other than when the accumulated value and preset are equal.

Comparison instructions are input instructions that test two values to determine if the instruction will be true or false. Comparison instructions include equal, not equal, less than, less than or equal, greater than, greater than or equal, masked comparison for equality, and the limit test, all to test one value against another. Most comparison instructions have two parameters, source A and source B. Typically, source A must be an address, while source B can be either a data file address or a constant. As an example, using an *Equal* instruction, source A could be a counter's accumulated value, and source B could be a constant, such as 10. When the counter accumulated value is equal to 10, the instruction will be true. On the other hand, if source B is an address, say N7:0, then the value stored in N7:0 is used in the comparison process.

LAB ONE

For this exercise, we will program a free-running timer to send a two-second pulse into a counter. The counter will provide a changing numerical value that will be used to evaluate the comparison instructions. We will view the comparison instructions as their source A values, represented by C5:0.ACC changes as the instructions become true or false.

1. _____ Open the Begin project.
2. _____ Create the ladder program as illustrated in Figure 17-11.
3. _____ Assign symbols to each input.
4. _____ Add rung comments as illustrated in Figure 17-11.
5. _____ When finished programming, download and save as compare. Go on-line with the project.
6. _____ As the program runs, observe the counter accumulated values change and note the true or false state of each of the instructions.
7. _____ Answer questions 8 through 15.

8. When is the *equal* instruction true? _____

9. When is the *not equal* instruction true? _____

10. As the counter increments, when is the *less than* instruction true? _____

11. When is the *less than or equal* instruction true? _____

12. When is the *greater than or equal* instruction true? _____

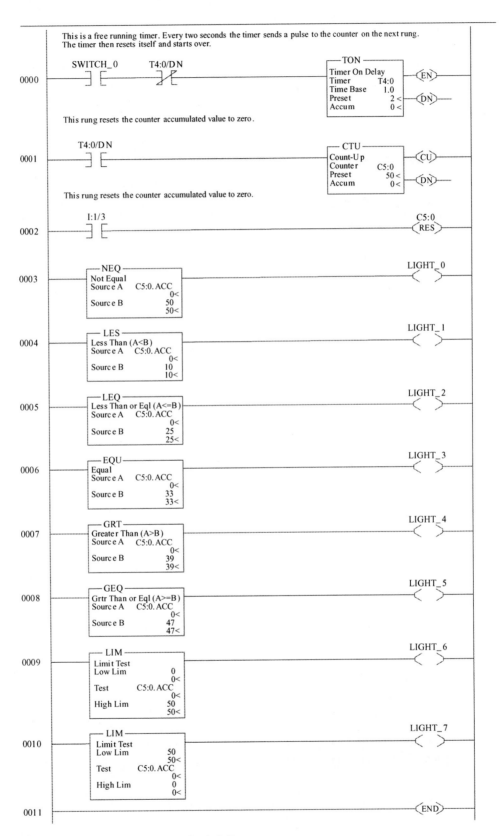

This is a free running timer. Every two seconds the timer sends a pulse to the counter on the next rung. The timer then resets itself and starts over.

0000

SWITCH_0   T4:0/D N

┤ ├   ┤/├

```
┌─ TON ──────────────────┐
│ Timer On Delay         │──(EN)
│ Timer        T4:0      │
│ Time Base     1.0      │
│ Preset          2 <    │──(DN)
│ Accum           0 <    │
└────────────────────────┘
```

This rung resets the counter accumulated value to zero.

0001

T4:0/D N

┤ ├

```
┌─ CTU ──────────────────┐
│ Count-U p              │──(CU)
│ Counter      C5:0      │
│ Preset         50 <    │
│ Accum           0 <    │──(DN)
└────────────────────────┘
```

This rung resets the counter accumulated value to zero.

0002

I:1/3

┤ ├

C5:0
(RES)

0003

```
┌─ NEQ ──────────────┐
│ Not Equal          │
│ Source A  C5:0. ACC│
│              0<    │
│ Source B    50     │
│             50<    │
└────────────────────┘
```

LIGHT_0
( )

0004

```
┌─ LES ──────────────┐
│ Less Than (A<B)    │
│ Source A  C5:0. ACC│
│              0<    │
│ Source B    10     │
│             10<    │
└────────────────────┘
```

LIGHT_1
( )

0005

```
┌─ LEQ ──────────────────┐
│ Less Than or Eql (A<=B)│
│ Source A  C5:0. ACC    │
│              0<        │
│ Source B    25         │
│             25<        │
└────────────────────────┘
```

LIGHT_2
( )

0006

```
┌─ EQU ──────────────┐
│ Equal              │
│ Source A  C5:0. ACC│
│              0<    │
│ Source B    33     │
│             33<    │
└────────────────────┘
```

LIGHT_3
( )

0007

```
┌─ GRT ──────────────┐
│ Greater Than (A>B) │
│ Source A  C5:0. ACC│
│              0<    │
│ Source B    39     │
│             39<    │
└────────────────────┘
```

LIGHT_4
( )

0008

```
┌─ GEQ ──────────────────┐
│ Grtr Than or Eql (A>=B)│
│ Source A  C5:0. ACC    │
│              0<        │
│ Source B    47         │
│             47<        │
└────────────────────────┘
```

LIGHT_5
( )

0009

```
┌─ LIM ──────────────┐
│ Limit Test         │
│ Low Lim        0   │
│                0<  │
│ Test    C5:0. ACC  │
│                0<  │
│ High Lim      50   │
│               50<  │
└────────────────────┘
```

LIGHT_6
( )

0010

```
┌─ LIM ──────────────┐
│ Limit Test         │
│ Low Lim       50   │
│               50<  │
│ Test    C5:0. ACC  │
│                0<  │
│ High Lim       0   │
│                0<  │
└────────────────────┘
```

LIGHT_7
( )

0011

(END)

Figure 17-11 Ladder program for Lab D.

13. When is the *greater than* instruction true? _____

14. Explain the operation of the *limit test* instruction on rung 9. _____

_____

15. Explain the operation of the *limit test* instruction on rung 10. _____

_____

LAB TWO: THE SCALE WITH PARAMETERS INSTRUCTION

The Scale with Parameters instruction (SCP) is a valuable instruction for scaling raw input data into a format acceptable for the application. As an example, an operator inputs a motor speed as 0 to 1750 RPM into an operator interface device such as a Panel View. This information is sent by way of a network communication cable into PLC memory. Figure 17-12 illustrates data flow from the Panel View operator interface through the SLC 500 processor and out to the variable-frequency drive.

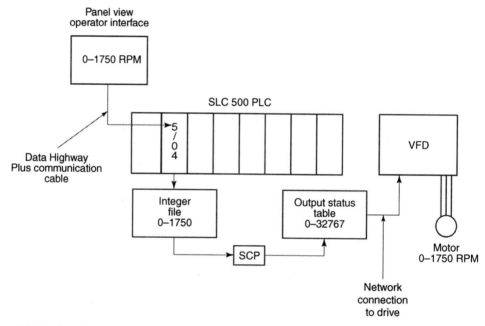

Figure 17-12 Data flow from a Panel View operator terminal into the PLC and out to a variable-frequency drive.

Since the SLC 500 has an intelligent backplane and assigns input status table words for only the input and output modules that actually exist in the PLC system, there are no extra words for inputs coming in from an operator interface device such as our Panel View operator interface terminal. The on and off status of the operator interface device screen objects such as push buttons and indicator lights, which are single-bit data, can be stored in the binary file. Numerical data such as motor speed can be stored in the integer file.

INSTRUCTION AVAILABILITY IN CONJUNCTION WITH YOUR PROCESSOR

The SCP instruction is operating system dependant. Go to instruction Help in your RSLogix 500 software and look in the upper left-hand corner of the screen under "use with processors."

Explain what is displayed. _____

_____

_____

_____

_____

LAB TWO

You must have a processor with an operating system that supports the SCP instruction before you can complete this lab exercise.

To demonstrate the SCP instruction, a timer will be used to increment from 0 to 1750 to simulate RPM. This is represented as (A) in Figure 17-13. The SCP instruction will be programmed with an input minimum of 0 and an input maximum of 1750 (B). The instruction will scale, or convert, the input data to the data format required by the variable-frequency drive. Since the drive is expecting a data range of 0 to 32767 representing 0 to 1750 RPM (0 to 60 Hz), the output minimum will be 0, while the output maximum will be 32767 (C).

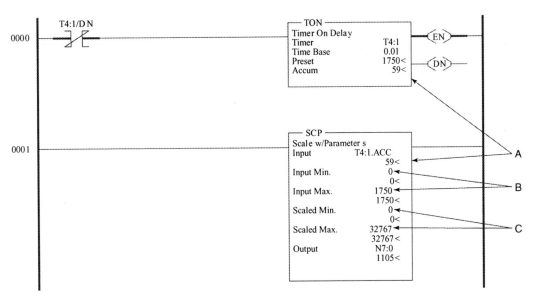

Figure 17-13   Timer simulating increasing motor speed for scaling by the SCP instruction.

1. _____ Program the rungs illustrated in Figure 17-13 to your compare program.
2. _____ When finished programming, download and save as comparison. Go on-line with the project.
3. _____ As the program runs, observe the timer accumulated values and how the SCP scales the input to the desired data format expected by the variable-frequency drive.
4. _____ Ask your instructor if the execution of your exercise needs to be checked off.
5. _____ Ask your instructor if this exercise is to be printed and handed in.

Analog Input Signal and the SCP Instruction

If you have an analog input on your PLC demo unit, modify your program using the SCP instruction to accept the analog input data.

1. _____ List the part number of your analog module. _____

2. _____ Record the input voltage or current from your field input device. _____

3. _____ Check the technical data on your analog manual to determine the data range for your particular module and voltage or current range. If you do not have access to the required technical information, use a personal computer to go to the Allen-Bradley internet site (http://www.ab.com). On the left side of the screen you should see General Resources.

   Click on Publications Library.

   Click on Manuals Online.

   Click on I/O.

   Click on Chassis-Based I/O.

   Click on 1746 SLC I/O. Look for the catalog number of the module.

   As an example, a 1746-NIO4V is a two-channel analog input and a two-channel analog voltage output module. Assuming a 0- to 10-volt input signal, the data range displayed in the input status table will be 0 to 32,767.

4. _____ Assuming you are using a 1746-NIO4V module with a pot sending a 0 to 10 Vdc input signal, open the input status file.

5. _____ Change the Radix of the data file to Decimal, refer to (A) in Figure 17-14.

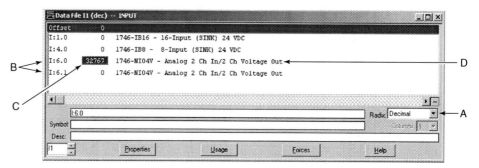

Figure 17-14   Input status table illustrating two-channel analog input module residing in slot 3.

6. _____ Refer to (B) in Figure 17-14. Notice the addressing format.

   Since this is a two-channel analog input module, two 16-bit words will be required to store the analog data. The address shows an input module in slots 1 and 4. The input status table for the module in slot 6 has been assigned two data words, I:6.0 and I:6.1. One word will represent each channel of the analog input module. Your slot number may be different depending on the slot in which your analog module resides. Also, if your analog module has more than two channels, there will be one word assigned for each channel.

7. _____ Turn the potentiometer to input 0 to 10 volts into the input module.

8. _____ Notice the value changing at (C). If your module inputs a 0 to 32767 signal, you should see the numbers change between 0 and 32767 as you turn the pot when in Decimal Radix.

9. _____ The Radix was changed because it is easier to monitor a decimal value representing the analog value.

10. _____ Notice Module Identification (D).

11. _____ For this example, the analog module has two input channels and two output channels. Open the output status file and view the two words created for the output channels of the module.

12. _____ With the information you have regarding the analog input module, program an SCP instruction to take the input information and scale it to 0 to 100 degrees. Store the scaled information in N7:0.

13. _____ When finished programming, download and save. Go on-line with the project.

14. _____ As the program runs, observe how the SCP scales the input to the desired data format and stores it in N7:0. Answer the Review Questions.

15. _____ Ask your instructor if your exercise needs to be checked off.

16. _____ Ask your instructor if this exercise is to be printed and handed in.

SCP REVIEW QUESTIONS

1. What did you program for the SCP input parameter? _____

2. How did you determine this? _____

   _____

3. What was the input minimum? _____

   _____

4. How did you determine this? _____

   _____

5. What was the input maximum? _____

   _____

6. What did you program as the scaled minimum? _____

7. What did you program as the scaled maximum? _____

8. There are small arrows pointing to the left next to each parameter. Explain what they represent. _____

   _____

   _____

9. If there were a four-channel analog input module residing in slot 7, list the addresses for each channel in the table in Figure 17-15.

ANALOG CHANNEL ADDRESSING	
Module Channel	**Input Address of Channel**
Channel 0	
Channel 1	
Channel 2	
Channel 3	

Figure 17-15   Channel addresses for question 9.

10. A 16-channel input module resides in slot 14. What would the address be for an analog device connected to channel 12? _____

11. When programming the SCP instruction, can a floating-point file be entered as an instruction parameter? _____

    A. Where would you go to find the answer? _____

    B. If floating-point parameters are permitted, explain how the instruction will treat its parameters if one parameter is entered as a floating point. _____

    _____

    _____

    _____

12. When entering the input max parameter, as an example, what options do you have regarding data types? _____

_____

_____

A. What is the difference between a data element and a constant? _____

_____

13. What is a long word or double-word? _____

_____

14. Are double-words available for use on the SLC 500 modular processors? _____

LAB THREE: SLC 500 SCALE DATA INSTRUCTION (SCL)

The Scale Data instruction (SCL) is used to scale raw input data to the PLC or output data to a field device. The Scale Data instruction is available on all MicroLogix and 5/02, 5/03, 5/04, and 5/05 modular processors. The Scale With Parameters instruction (SCP) is only available on the SLC 5/03–OS 302 and later processors, the SLC 5/04–OS 401 and later processors, the 5/05 modular processors, and the MicroLogix 1200 and 1500 PLC.

Let's look at the following example: A potentiometer ("pot") is inputting 0 to 10 volts into an analog input module. (The pot is used to change the speed command that goes to a variable frequency drive.) The drive speed information will be displayed on a PanelView Operator interface screen so the operator can see the speed value as they turn the pot.

Analog input signal is 0 to 10 Vdc.

Data displayed on the PanelView screen is 0 to 1750 RPM.

Raw data from the analog module is 0 to 32767.

Data from the PLC to the variable frequency drive via a network connection is 0 to 32767.

Figure 17-16 illustrates the SCL instruction from the RSLogix 500 software.

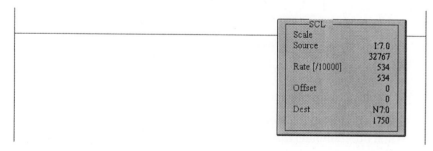

Figure 17-16 SLC 500 Scale Data instruction.

The SCL parameters are listed below:

Source:	The source must be a word address. Here we are looking at the analog channel I:7.0. The current raw analog input data is 32767.
Rate:	The rate is the slope. Enter a word address or a positive or negative integer that will be divided by 10,000.

Offset: Offset can be used as a correction factor if desired. This parameter can be either a constant value or the word address where the information will be found.

Destination: The destination is the word address where the rounded, scaled result will be stored.

Application Example

Let's assume we have a pot inputting 0–10 Vdc into channel zero of the analog card in slot 7 of our chassis. This is address I:7.0. This particular card inputs raw data of 0–32767 from a 0–10 volt signal. We need to scale this raw data from 0–32767 to 0–1750 in order to represent the desired motor RPM. This information will be stored in the destination N7:0 for display on an operator interface screen so the operator can see what speed has been selected.

The scaled value = (input value × rate) + offset

$$1750 = (32767 \times \text{rate}) + \text{offset}$$

The rate = (scaled maximum − scaled minimum)/(input maximum − input minimum)

1750 − 0/32767 − 0 = .053407

To get the rate parameter, multiply .053407 by 10000

The rate equals 534.07. Enter a whole number into the rate parameter. Floating point will not be accepted.

The math:

Scaled value = 32767 × (534/10000)

Scaled value = 32767 × .0534

Scaled value = 1750, which will be stored in the destination N7:0

If the actual scaled value came up as 1745, an offset of 5 will bring it up to 1750.

LAB EXERCISE

Use the SCL instruction to scale the raw input data from a 1746-FIO4I fast-response two-channel analog input and output module. The input signal is from address I:6.3 and is 4 to 20 mA. Refer to Figure 17-17. Specifications for the SCL instruction are listed below:

4–20 mA = 409−2047

Offset = 0

Scale from 0 to 100%

Store scaled data in N10:6 for use in our program.
Show your calculations below.

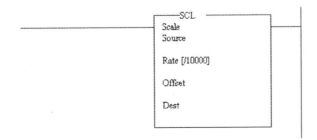

Figure 17-17   Fill in the SCL instruction for your answer.

LAB E: Applying Hexadecimal Numbers in PLC Masking Applications

OBJECTIVES

Upon completion of this laboratory exercise, you should be able to:

- understand hexadecimal masking as used with ControlLogix, PLC 5, and SLC 500 family instructions
- review the hexadecimal number system
- develop an RSLogix 500 program incorporating the Masked Move instruction

INTRODUCTION

Most modern PLCs have instructions that allow manipulation of data in different ways. Some PLC data manipulation instructions use masks, or filters, to block out unwanted data from passing from one location to another. A mask is simply a method to control data flow. Think of a mask as a line of sixteen doors, one door for each bit in a 16-bit word. If the door is open, data can pass; if the door is closed, data is held back, or masked out. Considering that a single hexadecimal number is equivalent to four bits, a four-digit hexadecimal number, sixteen bits, could act as a mechanism to control data flow bit by bit.

Current Allen Bradley PLCs use hexadecimal masking for instructions such as the Masked Move, Sequencer Output, Masked Equal, Immediate Input with Mask, and Immediate Output with Mask. The PLC 5 and SLC 500 family of PLCs use the same Masked Move Instruction. Since the PLC 5 and SLC PLCs are 16-bit computers, their instructions and masks are also 16 bits. The newer ControlLogix is a 32-bit computer that uses the same instruction set as the PLC 5. Even though ControlLogix instructions use 32-bit hexadecimal masking, in most cases the instructions work the same as the PLC 5 and SLC 500. In this section, we will investigate the SLC 500 Masked Move instruction and how a hexadecimal mask is used in conjunction with moving selected data from one memory location through a hexadecimal mask to another.

The Move instruction is used to move a copy of information contained in a source, which is typically a data table location, to another user-specified data table location. The 16-bit data word in the starting location is called the source word, while the target location to which data is being moved is called the destination. An example of moving data from one location to another would be moving a value from an integer table into a timer or counter instruction. The data could specify how long a mixer is to mix or how many counts of a specific product signify a full case. Each product manufactured may have different mixing parameters or differing counts signifying a full case. The correct time or count value would need to be moved into the user ladder diagram timer or counter parameters as dictated by which product is being manufactured. In applications such as these, the entire 16-bit word represents a whole number that is a time or count value; the entire data word would have to be moved.

In some applications, only selected bits of the 16-bit data word need to be moved or separated from adjacent bits. The masked move instruction uses a hexadecimal mask to mask, or filter out, undesired bits in a 16-bit data word. The mask restricts data from being moved from the source 16-bit word to the destination.

Hexadecimal masks are used because it is easier to enter a four-digit hexadecimal number than sixteen bits of binary information. The instructions on your ladder program will prompt you to enter a hexadecimal value for your mask instruction parameter. When working with hexadecimal masks, think of the actual mask in its binary 1s and 0s state. Each mask binary bit will pair up with its associated binary bit in the source word to determine if the source data will pass. This is important, as the bit value of each of the sixteen bits will determine if data is to pass. Think of the hexadecimal mask as a set of sixteen doors, one door for each bit of the source to pass through to the destination. If the door is open for a bit, the bit will be allowed to pass through to the destination. If the door is closed, the source bit is not allowed to pass. When the source bit is not allowed to pass, the destination bit remains in its same logical state.

Before we look at the masked move instruction, let's review the basic *move* instruction. The move instruction moves a *copy* of the data contained in the source location to the designated destination. So the term *move instruction* is a bit deceiving, as it implies moving data *from* one location to another. In fact, the move instruction places a copy of the data in the designated destination, leaving the original data in place at the source location.

The move instruction is an output instruction with two parameters:

A. The *source* is an address where data is stored or a constant entered into the source parameter of the instruction.
B. The *destination* is the address where a copy of the data is to be placed, or "moved."

If a copy of the data is to be placed only once, a one-shot instruction should be programmed ahead of the move instruction. Figure 17-18 illustrates a move instruction used in conjunction with a one-shot.

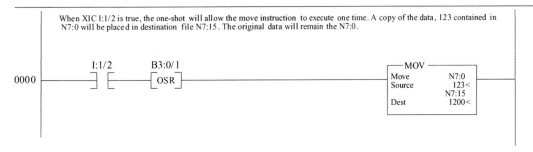

When XIC I:1/2 is true, the one-shot will allow the move instruction to execute one time. A copy of the data, 123 contained in N7:0 will be placed in destination file N7:15. The original data will remain the N7:0.

```
                I:1/2      B3:0/1                                      ┌──── MOV ────────┐
0000    ────────┤ ├────────┤ ├────[ OSR ]─────────────────────────────┤ Move            │
                                                                       │ Source    N7:0  │
                                                                       │           123<  │
                                                                       │           N7:15 │
                                                                       │ Dest      1200< │
                                                                       └─────────────────┘
```

Figure 17-18   Move instruction used in conjunction with a one-shot instruction.

Evaluating the rung, each time I:1/2 becomes true, the one-shot will trigger the move instruction. The move instruction (MOV) will place a copy of the data contained in N7:0 into N:15. The data 123 which is contained in N7:0 will also be found in N7:15.

MASKED MOVE INSTRUCTION

In some instances we may not desire to move all sixteen source bits to a specific destination. If only the lower eight bits of the source word are to move to the destination, we need a way to block, or mask out, the bits we do not desire to move. This can be accomplished by passing the bits through a controlling mechanism which allows us to control which bits pass through to the destination. This controlling mechanism is called a *mask*.

The *masked move instruction* moves a *copy* of the data contained in the source location through a *hexadecimal mask* to the designated destination. A copy of the original data is moved through a user-designated hexadecimal mask parameter to the designated destination; the original data remains in the source location.

The masked move instruction is an output instruction with three parameters:

A. The *source* is an address where data is stored or a constant entered into the source parameter of the instruction.

B. The *mask* is the address where the mask will be found or a constant hex value.

C. The *destination* is the address where a copy of the data is to be placed or moved.

The masked move instruction will move data to the designated destination for each scan the instruction is true. If a copy of the data is to be moved only once, a one-shot instruction should be programmed ahead of the move instruction. Figure 17-19 illustrates a masked move instruction used in conjunction with a one-shot.

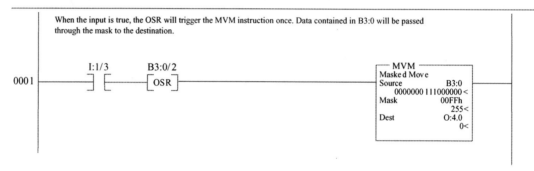

Figure 17-19   The Masked Move instruction.

Evaluating the rung, each time I:1/3 becomes true, the one-shot will trigger the masked move instruction. The masked move instruction (MVM) will move a copy of the data contained in B3:0 to O:4.0 through the hex value designated as the mask.

UNDERSTANDING HEXADECIMAL NUMBERS

Even though hexadecimal (hex) may seem intimidating at first, the only thing you really need to know about hex is illustrated in the table in Figure 17-20.

HEXADECIMAL CORRESPONDING BIT PATTERNS			
0	0000	8	1000
1	0001	9	1001
2	0010	A	1010
3	0011	B	1011
4	0100	C	1100
5	0101	D	1101
6	0110	E	1110
7	0111	F	1111

Figure 17-20   Hex and binary bit pattern correlations.

The object behind hex is to be able to use all 16-bit patterns represented by four bits. As you remember from our study of BCD, 4-bit codes between 10 and 15 were invalid BCD codes. In many cases, if you were to attempt to convert an invalid BCD code in an SLC 500 program, the processor would fault. Hex is an extension of BCD, thus allowing the use of the invalid BCD codes 10 through 15. The problem when hex was being developed was that a method had to be found to represent the values 10 through 15 using only one character. This was in the days of typewriters and teletype machines. Creating new characters to represent these new codes would dictate redesigning all keyboards. This was an unacceptable option. The simplest solution was to use the first letters of the alphabet to represent the values 10 through 15. As a result, the letters A through F were used to represent these new codes. See Figure 17-21.

Decimal Value	Hex Character	Binary Equivalent
10	A	1010
11	B	1011
12	C	1100
13	D	1101
14	E	1110
15	F	1111

Figure 17-21   Decimal values and their hex and binary equivalents.

For you to understand and use hex masking all you need is in the table in Figure 17-20. The mechanics of masking are really quite simple. You will take a source data word where some of the sixteen bits need to be filtered out or stopped from moving from the source to the destination. As an example, a BCD value, 1234, input using thumbwheels, actually represents two separate pieces of data. The value 12 identifies the product size or color, while the 34 means that a quantity of 34 pieces needs to be produced. As input into the PLC, it is the value 1,234. This information needs to be separated into two groups of data and stored in separate integer file locations. Assume the 12 needs to be stored in N10:15, and the 34 is to be stored in N10:16. With the data separated, the PLC has information it can work with. As products are made, the PLC will look at N10:15 to determine size and at N10:16 to find the number of pieces to be produced. A move instruction is used to move the value 34 into a counter preset value. A mechanism is needed to separate the two values. This is where the masked move instruction comes into the picture. One masked move instruction will be used to allow the 34 to move from the source to the destination. The value 12 will be masked out, or not allowed to move. A second masked move instruction is programmed to move the 12 from the source to the destination, thus masking out the 34. The first masked move instruction is straightforward; however, the second instruction has a problem that must be corrected. Since the 12 is in the upper byte of the 16-bit word, after executing the MVM instruction, the value 12 will still be in the upper byte. However, there will be zeros in the lower byte. The resulting value will actually be 1200. This value needs additional conversion before it is correct. Do some research and see how the SWP instruction could help here.

Applying Hexadecimal Numbers and Masks with SLC 500 Programming Instructions

The SLC 500 mask parameter uses either a four-character hexadecimal numerical code as the mask word or the address where the mask word will be found. This hex word, used in its binary equivalent form, is the mask through which the masked move instruction will move the data. If the mask bits are *set*, or a logical one, data will pass through the mask, while with bits that are *reset*, a logical zero, the mask will restrict data from passing. Let's look at an example.

Example One
Figure 17-22 illustrates a source word 0110 1111 1010 1100. The lower eight bits need to pass through the mask to the destination. The upper eight bits of the source word are to be blocked by the mask. As a result, the lower eight source bits will be reflected in the destination word, while the upper eight destination bits will remain unchanged. Figure 17-22 identifies bits to pass with a Y for yes and an N for no.

0110	1111	1010	1100	Source word
NNNN	NNNN	YYYY	YYYY	Bit to pass through to destination?
Bits do not change		1010	1100	Destination word

Figure 17-22   The lower eight source bits will be allowed to pass through the mask.

To set up the mask to pass the correct data, set a 1 in the mask bit position where each source data bit is to pass. This is illustrated in Figure 17-23.

0110	1111	1010	1100	Source word
0000	0000	1111	1111	Bit to pass through to destination?
Bits do not change		1010	1100	Destination word

Figure 17-23   Bits allowed to pass move through 1s in Hex mask.

Next, convert bit information to hex, as illustrated in Figure 17-24. The converted hex value will be entered into the mask parameter of the masked move instruction.

The mask bit pattern	0000	0000	1111	1111
Hex value of	0	0	F	F

Figure 17-24   Mask bits translated to Hex.

Example Two

The next example will illustrate how to pass only selected bits in the lower byte of the source data word to the destination. Note that the Xs in the destination word mean these bits do not change from their original contents. See Figure 17-25.

0110	1111	1010	1100	Source word
NNNN	NNNN	YYNN	NNYY	Bit to pass through to destination?
Bits do not change		10XX	XX00	Destination word

Figure 17-25   Ys indicate which bits are to pass through the mask.

To set up the mask so as to pass the correct data, you simply set a 1 in the mask bit position where each source data bit is to pass. This is illustrated in Figure 17-26.

0110	1111	1010	1100	Source word
0000	0000	1100	0011	Bit to pass through to destination?
Bits do not change		10XX	XX00	Destination word

Figure 17-26   Ys from Figure 17-25 changed to 1s.

Convert bit format to hex; this will be entered into the mask parameter of the masked move instruction, as illustrated in Figure 17-27.

The mask bit pattern	0000	0000	1100	0011
Hex value of	0	0	C	3

Figure 17-27   Mask binary bit pattern converted to Hex.

LAB ONE

We will now develop a ladder rung for each of the previous examples. Our first example of the masked move instruction started with the source word of 0110 1111 1010 1100 and moved it through a 00FF mask. Figure 17-28 illustrates the program rung we will be creating.

1. _____ Open your Begin project.
2. _____ Open Binary file B3 and enter 0110 1111 1010 1100 into B3:0.

Figure 17-28   Masked Move ladder rung.

3. _____ Create the ladder rung as illustrated in Figure 17-28.
4. _____ Save your program as Masked Move Lab.

Testing Your Program

Fill in the table in Figure 17-29 with the source and mask information. Notice that the MVM source parameter is the binary value 0110 1111 1010 1100 and the mask is in hex.

1. _____ Record the binary bit pattern in the source area of Figure 17-29.
2. _____ Record the mask in the mask area of Figure 17-29.

				Source word
				Mask
				Expected destination word

Figure 17-29   Expected result from executing Masked Move instruction.

3. _____ Fill in the expected destination word in Figure 17-29.
4. _____ Energize input I:1/3. The MVM instruction will be executed once.
5. _____ What output LEDs on the output module in slot 2 are energized?
6. _____ Does the output bit pattern correspond to the expected values you entered in Figure 17-29?

Viewing the Output Status Table Bit Pattern

Let's look at the output status table bit pattern and compare it to the destination word from Figure 17-29.

1. _____ Open the output data file and view the bit pattern in word O:2.0.

LAB TWO

We will now develop a ladder rung for rung two of the previous example. This example of the masked move instruction source word is 0110 1111 1010 1100 and is moved through a 00C3 mask. Figure 17-30 illustrates the program rung we will be adding to our program.

We will be adding logic illustrated in Figure 17-30 to the current ladder program.

1. _____ Program the second rung as illustrated in Figure 17-30 using the same basic procedure you used to program rung one.
2. _____ Record the binary bit pattern of the source in Figure 17-31.
3. _____ Record the mask in Figure 17-31.
4. _____ Fill in the expected destination word in Figure 17-31.
5. _____ Energize input I:1/4. The MVM instruction will be executed once.
6. _____ What output LEDs on the output module in slot 4 are energized?

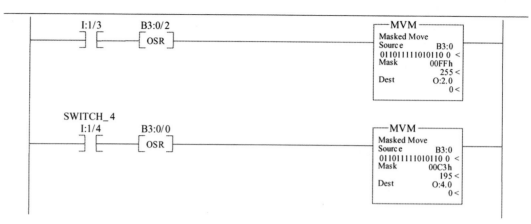

Figure 17-30 Masked Move instruction rung for exercise.

				Source word
				Mask
				Expected destination word

Figure 17-31 Expected result after executing Masked Move instruction.

7. _____ Does the output bit pattern correspond to the expected values you entered in Figure 17-31?

8. _____ Look at the output status table bit pattern and compare it to the destination word from Figure 17-30.

Changing Source Data for the MVM on Rung Two

This exercise will edit the MVM instruction on the second rung and change the source address from B3:0 to B3:1.

1. _____ Edit the second rung's MVM instruction so that the source address is B3:1.
2. _____ Open Binary file B3 and enter 0000 0000 1111 1111 into B3:1.
3. _____ Close the B3 data file when finished.
4. _____ Record the binary bit pattern of the source in Figure 17-32.
5. _____ Record the mask in Figure 17-32.
6. _____ Fill in the expected destination word in Figure 17-32.
7. _____ Download, put processor in run mode, and go on-line.
8. _____ Energize I:1/4. Do the PLC output module LEDs correspond to the expected result?
9. _____ Go off-line with the processor.

				Source word
				Mask
				Expected destination word

Figure 17-32 Expected result after executing rung 2:1.

Changing Masks

This exercise will edit the MVM instruction on the first rung and change the mask. Use the same procedure you used to edit your ladder program rung to change the source address; but, rather than changing the source address, change the mask as specified below. As your program runs, observe the output module's LEDs and verify that the output data word corresponds to the expected destination word for the MVM instruction.

Practice I

1. _____ Edit the first rung so the mask is stored in B3:10 and is now ABCD.
2. _____ Record the binary bit pattern of the source in Figure 17-33.
3. _____ Record the mask in Figure 17-33.
4. _____ Fill in the expected destination word in Figure 17-33.
5. _____ Download, put processor in run mode, and go on-line.
6. _____ Energize I:1/3. Do the PLC output module LEDs correspond to the expected result?
7. _____ Go off-line.

				Source word
				Mask
				Expected destination word

Figure 17-33   Expected result after executing rung 0001.

Practice II

1. _____ Edit the first rung so the mask is now 0F0F.
2. _____ Record the binary bit pattern of the source in Figure 17-34.
3. _____ Record the mask in Figure 17-34.
4. _____ Fill in the expected destination word in Figure 17-34.
5. _____ Download, put processor in run mode, and go on-line.
6. _____ Energize I:1/3. Do the PLC output module LEDs correspond to the expected result?

				Source word
				Mask
				Expected destination word

Figure 17-34   Expected result after executing rung.

Practice III

1. _____ Edit the first rung so the mask is now FF00.
2. _____ Record the binary bit pattern of the source in Figure 17-35.
3. _____ Record the mask in Figure 17-35.
4. _____ Fill in the expected destination word in Figure 17-35.
5. _____ Download, put processor in run mode, and go on-line.
6. _____ Energize I:1/3. Do the PLC output module LEDs correspond to the expected result?

				Source word
				Mask
				Expected destination word

Figure 17-35   Expected result after executing rung.

Practice IV

1. _____ Edit the first rung so the mask is now address N7:3.
2. _____ Monitor the data table N7. Change radix to hex. Enter DCBA as mask data in N7:3.
3. _____ Record the binary bit pattern of the source in Figure 17-36.
4. _____ Record the mask in Figure 17-36.
5. _____ Fill in the expected destination word in Figure 17-36.
6. _____ Download, put processor in run mode, and go on-line.
7. _____ Energize I:1/3. Do the PLC output module LEDs correspond to the expected result?

				Source word
				Mask
				Expected destination word

Figure 17-36 Expected result after executing rung.

LAB TWO EXERCISES

1. Fill in the destination word in Figure 17-37. Assume the beginning destination word is 0000 0000 0000 0000 unless instructed otherwise.

1010	1010	1010	1010	Source word
0000	1111	0000	1111	Mask
				Destination word

Figure 17-37 Expected destination word after executing instruction.

The Hex mask is _____.

2. Fill in the mask and destination word in Figure 17-38 as if the hex mask is 00FF.

1110	1001	0011	0001	Source word
				Mask
				Destination word

Figure 17-38 Expected destination word after executing instruction.

3. Fill in the source word in Figure 17-39 as if the hex mask is ABCD.

				Source word
A	B	C	D	Mask
1000	0011	0010	1101	Destination word

Figure 17-39 Source word needed to produce desired Masked Move instruction execution results.

4. Fill in hex mask in Figure 17-40 if:

1111	1000	1010	1101	Source word
				Mask
1101	0000	0000	1000	Destination word

Figure 17-40 Mask needed to produce desired Masked Move instruction execution results.

5. Fill in mask and destination in Figure 17-41 as if we wanted to mask out the upper byte.

1101	0010	0000	0111	Source word
				Mask
				Destination word

Figure 17-41 Expected destination word after executing instruction.

6. Fill in mask and destination in Figure 17-42 as if we wanted to only allow bits 6 through 13 to pass.

1110	0100	1100	0001	Source word
				Mask
				Destination word

Figure 17-42 Expected destination word after executing instruction.

7. Your source word is 1101111011010111. The mask is F0B4. What is the destination word?

_____

8. Your destination word is 1011011011011010. The mask is 1C2B. What is the source word?

_____

9. Create ladder rungs on your current masked move program so when:

   A. Switch one is true, mask 00FFh is moved into the masked move instruction on current rung 0001.
   B. Switch one is false, mask FF00h is moved into the masked move instruction on current rung 0001.

10. Drive Interface Application

Currently an SLC 500 PLC is controlling the manufacturing of your product. You are asked to install an Allen-Bradley Bulletin 160 AC variable-frequency drive so an SLC 500 PLC can send a 3-bit output signal to the drive to control the motor speed. This drive comes in two versions, preset or analog follower. For this application, the preset version will be used. The preset version accepts a combination of three inputs to TB3 to command the speed at which the drive is to run the motor. Three switches can provide up to eight different on/off input combinations to the drive. Figure 17-43 illustrates three switches connected to TB3 to show speed reference input.

   Installing and wiring the drive, we discover there is only one output module with output points available. Screw terminals 6, 7, and 8 are the only unused output points on the output

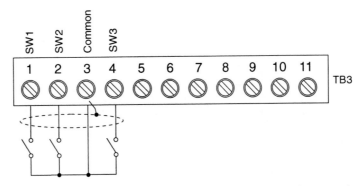

Figure 17-43 Three input signals into TB3 communicate to the drive what speed to run the motor.

module residing in slot 2 of the chassis. All other outputs control other parts of the production process and have no direct correlation to our drive wiring. Figure 17-44 illustrates the output module and the speed control wiring to the drive's terminal block. The drawing has been simplified to illustrate the point and is not necessarily technically correct.

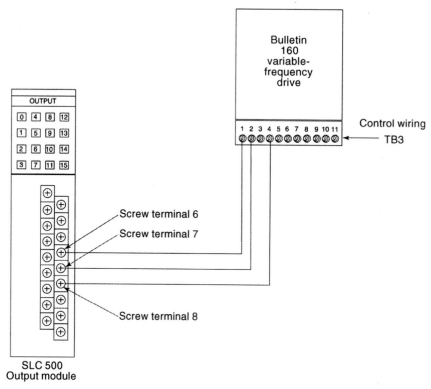

Figure 17-44 Simplified overview of PLC output module wiring to drive's control wiring terminal block TB3.

You have to find a way to send a 3-bit signal out screw terminals 6, 7, and 8 to control the drive without affecting the other 13 output points on the module.

When working with PLCs, we use instructions; drives use parameters. Instructions and parameters are basically the same thing with different names. The user will program the drive by entering the desired preset frequencies into parameters 61 through 68; the commanded speed value will be in Hz. Not all preset frequency parameters need to be used for a specific application. For the drive to determine at what speed to control the motor, the drive will look in the currently selected parameter. The table in Figure 17-45 lists the parameters, preset frequency,

Parameter	Preset Frequency	Default	TB3-4	TB3-2	TB3-1
61	Preset 0	3.0 Hz	0	0	0
62	Preset 1	20.0 Hz	0	0	1
63	Preset 2	30.0 Hz	0	1	0
64	Preset 3	40.0 Hz	0	1	1
65	Preset 4	45.0 Hz	1	0	0
66	Preset 5	50.0 Hz	1	0	1
67	Preset 6	55.0 Hz	1	1	0
68	Preset 7	60.0 Hz	1	1	1

Figure 17-45 Drive parameters, their associated speed in Hz, and the signals sent to TB3.

default frequency, and bit patterns required to select the desired preset frequency. The default speeds listed in the table are values programmed at the factory as a starting point. Parameter values can be modified by the user. Parameters are listed as Hz or hertz. A drive's output to a motor is commonly referred to in Hz. Typically, 0 to 60 Hz equals 0 to 1750 RPM motor speed.

The binary bit patterns associated with each parameter will select the preset speed command the drive will receive. These signals will come ultimately from the RSLogix 500 program by way of the output status table, through the output module, to the drive. The bit patterns will be seen by the drive at the three screw terminals on TB3. Referring to the table, if TB3-1 = 0, TB3-2 = 1, and TB3-4 = 0, the preset frequency value programmed in parameter 63, currently 30.0 Hz, will be the commanded frequency. When selected, the programmed value in the parameter sets the frequency the drive outputs.

Drive Parameter Programming

For our application, the motor will need to run at three different speeds during the manufacturing process. The three speeds are 45 Hz, 52 Hz, and 60 Hz. These will be programmed into drive parameters 61, 62, and 63.

For this exercise, we are only considering PLC programming to correctly control the speed reference bits sent from the RSLogix ladder as an output to the drive. We will not worry about wiring or drive programming.

NOTE: If using a MicroLogix 1000, use output screw terminals 6, 7, and 8. If not available, use terminals 2, 3, and 4.

As you consider your programming options, answer the following questions:

1. _____ Explain how you will control the 3-bit speed signal flow in the PLC program to the output module and on to the drive. _____

_____

_____

_____

2. _____ What instruction(s) will you use? _____

_____

3. _____ Sketch the instruction(s) used and provide all programming parameters.

4. _____ Explain in detail why you selected this instruction(s). _____

_____

_____

5. _____ Define each instruction parameter and explain why you programmed it as you did.

_____

_____

_____

_____

REVIEW QUESTIONS

Refer to Figure 17-46 as you answer the following questions:

1. Explain how the LIM instruction on rung 0 works. _____

_____

2. When will the LIM instruction on rung 1 be true? _____

_____

3. Describe the difference between the LIM instruction on rung 0 and 1. _____

_____

_____

4. What is the function of the EQU instruction on rung 2? _____

_____

  A. What is source A? _____
  B. If the value in C5:0.ACC is 75, will the instruction be true or false? _____
  C. What value(s) in C5:0.ACC will make the instruction false? _____

5. Describe the operation of the NEQ on rung 3. _____

_____

_____

  A. What value is currently in N7:50? _____
  B. How do you know this? _____
  C. Is the instruction currently true or false? _____

6. Explain how the LES instruction operates on rung 4. _____

_____

_____

  A. Explain the address in source A. _____
  B. What is the current value in source A? _____
  C. When will the instruction be true? _____
  D. Is the instruction currently true? _____

7. How does the GRT instruction work on rung 5? _____

_____

_____

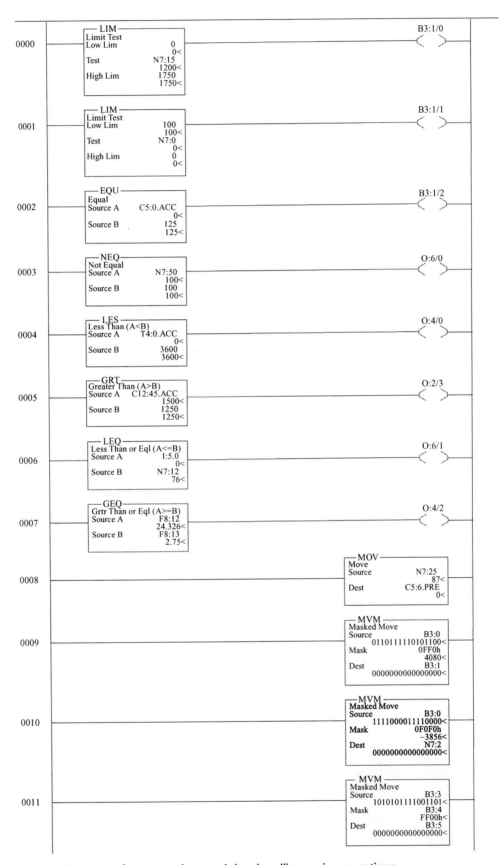

Figure 17-46 Ladder rungs for comparison and data handling review questions.

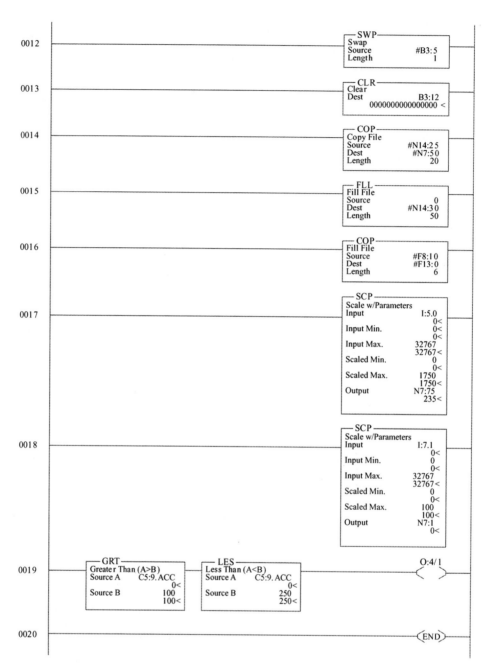

Figure 17-46 *(continued)*

A. When is the instruction true? _____

B. Is the instruction currently true? _____

8. Describe the operation of the LEQ instruction on rung 6. _____

_____

_____

A. Identify the address in source A. _____

B. When is the instruction true? _____

C. Is the instruction currently true? _____

9. Describe the operation of the GEQ instruction on rung 7. _____
   _____
   _____

   A. Identify the address in source A. _____

   B. When is the instruction true? _____

   C. Is the instruction currently true? _____

10. Explain how the MOV instruction operates on rung 8. _____
   _____
   _____

   A. What value is currently in N7:25? _____

   B. Define "destination address." _____

   C. Explain how you might use this instruction as programmed. _____
   _____

11. Describe the operation of the MVM instruction on rung 9. _____
   _____
   _____

   A. What kind of file is the source? _____

   B. Explain how the mask works. _____
   _____

   C. What is the binary bit pattern of the mask? _____

   D. Why is the far right character an "h" in the mask? _____

   E. Upon execution of this instruction, what will be stored in B3:1? ____
   _____

12. Answer the following referring to the MVM instruction on rung 10.

   A. Why is the mask value displaying a negative value? _____
   _____

   B. What is the binary bit pattern of the mask? _____

   C. Upon execution of this instruction, what will be stored in the destination? _____
   _____

13. Answer the following referring to the MVM instruction rung 11.

   A. What is different about this mask? _____

   B. Explain how this mask works. _____
   _____

   C. What is the binary bit pattern of the mask? _____

   D. Upon execution of this instruction, what will be stored in B3:5? ____
   _____

14. Explain the operation of the SWP instruction on rung 12. _____

_____

_____

A. If B3:5 contained a value of 00F0h before instruction execution, what would be the result of executing the instruction? _____

B. What does the # signify before the source parameter? _____

C. Explain the length parameter. _____

_____

D. What is the maximum value that can be programmed in the length parameter? If you do not know, where would you look to find this? _____

_____

15. Describe the function of the CLR instruction on rung 13. _____

_____

_____

A. Provide an example of where this might be used. _____

16. Explain how the COP instruction operates on rung 14. _____

_____

_____

_____

A. The source file is N14:25. What elements will be copied? _____

_____

B. When is there a # in front of the source and destination parameters? _____

_____

C. Define the length parameter. _____

_____

D. How many total elements can be copied with the COP instruction? _____

E. Is it possible to copy non-consecutive elements with one copy instruction? _____

17. Assume each day's production data was stored starting at N14:30. Every morning at 6:55 a.m. the prior day's production data is sent across the Data Highway Plus network to the Production Supervisor's personal computer Excel Spreadsheet. The first shift starts at 7:00 a.m. Explain how the FLL instruction on rung 15 could fit into this situation.

_____

_____

_____

_____

_____

Do a little research and explain how the addresses S:39-S:37 and S:40-42 could be used on your ladder in conjunction with this application. Assume you have a 5/03 or later processor.

A. Why does the FLL source not have a # programmed? _____

_____

B. Can the FLL source be other than a constant as illustrated in rung 15? _____

C. What data file contains addresses starting with S? _____

D. Define each of the following addresses:

S:42   _____

S:41   _____

S:40   _____

S:39   _____

S:38   _____

S.37   _____

18. In an unrelated application of the FLL instruction, do a little research and explain how the S:1/15 bit could be used on the first rung of your ladder program in conjunction with the FLL instruction to clear out old data after maintenance has been performed on the machine and the PLC is returned to run mode. _____

_____

_____

_____

_____

_____

19. Refer to the COP instruction on rung 16 and explain the difference between an F-type file and the N-type file used in the COP instruction on rung 14. _____

_____

_____

_____

_____

20. Refer to the SCP instruction on rung 17.

A. When we state "Instruction is operating system dependant," what do we mean? _____

_____

_____

B. What processors can use this instruction? Be specific. _____

_____

C. Where are you going to find this information? _____

D. Can the MicroLogix 1000 use this instruction? _____

21. List and define each SCP Parameter below.

22. We have a Panel View operator terminal where the operator can input the desired motor speed. The motor operating speed range is 0 to 1750 RPM. The motor speed is sent from the Panel View to the SLC 500 processor via a network. The 0 to 1750 speed reference is stored at address N10:23. The PLC program needs to verify that the speed is within the proper operating range. Over-speeding the motor is not allowed. If the desired speed data is within the operating range, it is output to N10:24. The variable-speed drive is expecting to see data in the format of 0 to 32767 equals 0 to 60 Hz (0–60 Hz = 0–1750 RPM motor speed). Sketch the ladder logic you would create for this application. Fill in the proper data for all instruction parameters.

23. Refer to rung 18 for this application question. The operator turns a potentiometer that inputs a 0- to 10-volt signal to represent the desired temperature of 0 to 100 degrees. The input signal of 0 to 10 volts is converted to 0 to 32767 by the input module.

    A. What type of input module will be used? _____

    B. Explain the correlation of the application and the SCP instruction. Identify how each instruction parameter fits into the application. _____

    _____

    _____

    _____

24. Refer to rung 19.

    A. Explain how rung 19 operates. _____

    _____

    _____

    B. Is there a better way to accomplish this? _____

    _____

25. A variable-frequency drive speed feedback signal to the PLC processor is in the format 0 to 32767 equaling 0 to 60 Hz. Assume the PLC will send the speed value on to an operator interface device such as a Panel View. The Panel View is to display the speed value as 0 to 1750 RPM for the operator to view.

A. Explain how this application correlates to the SCP instruction on rung 17. Define how each parameter fits into the picture. _____

_____

_____

_____

_____

B. Assume the speed value also needs to be displayed on a BCD display device. Refer to your text and describe how you will accomplish this on your ladder rung. Assume an output address of O:6.0. Be very exacting in your explanation. _____

_____

_____

_____

_____

_____

_____

26. Figure 17-47 shows three rungs of ladder logic from ControlLogix and its RSLogix 5000 software. Your task is to determine the destination tag after each Masked Move instruction is executed. The Masked Move instruction works the same as the 16-bit versions we have been working with for the PLC 5 and SLC 500. However, there are three differences.

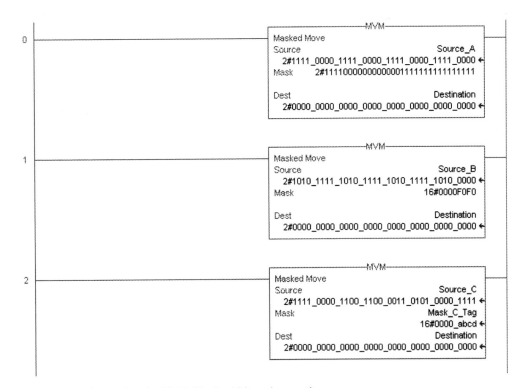

Figure 17-47   ControlLogix 32-bit Masked Move instructions.

A. The ControlLogix is a name-, or tag-based PLC, so the source and destination will be names (tags) rather than file addresses.
B. The source, mask, and destination will each be 32 bits wide.
C. The radix of information in ControlLogix is represented a bit differently. 16# signifies that the information is hexadecimal. 8# is octal; 2# means the data is binary. The lack of a designator indicates that the information is decimal.

Using the same principles for determining the destination after an SLC or PLC 5 Masked Move instruction is executed, determine the destinations for the ControlLogix Masked Move instructions in Figure 17-47. As an example, the rung 0 source tag is "Source A." The blue arrow directly below the tag is 2# 1111_0000_1111_0000_1111_0000_1111_0000. The 2# identifies this tag as binary data. The 32 bits of data follow to the right of the radix designator. Rung 0 has a binary mask whereas rungs 1 and 2 have Hexadecimal masks. The starting value of all destination tags is directly below the destination tag to the left of the blue arrow. Starting destination tags are all zeros.

A. _____ After the MVM instruction on rung 0 is executed, the destination tag

will be _____.

B. _____ After the MVM instruction on rung 0 is executed, the destination tag

will be _____.

C. _____ After the MVM instruction on rung 0 is executed, the destination tag

will be _____.

The Sequencer Instruction

OBJECTIVES

Upon completion of this laboratory exercise, you should be able to:

- understand hexadecimal masking as used with SLC 500 family sequencer instruction
- program the sequencer instruction
- create the sequencer's data table file
- understand how Control File R6 is used in conjunction with the sequencer instruction

INTRODUCTION

The *sequencer* instruction has become one of the workhorse features of the PLC. The sequencer simply controls a predetermined sequence of events, for example, the control of a pallet stretch wrap machine. Each step of the pallet wrapping routine is controlled by preprogrammed sequencer steps entered into the PLC user program. Each step of the sequence performs a predetermined task, such as:

1. Determine if adequate stretch wrap film is available.
2. Sense presence of a pallet to be wrapped.
3. Move pallet to proper position for wrapping.
4. Perform the wrapping process.
5. PLC is programmed to remember stretch wrap patterns. Plastic wrap is staggered to improve transportability of the loaded pallet. Each staggered wrapping pass is a sequence in itself.
6. When the wrapping process is completed, the pallet is sequenced to exit the conveyor section to be loaded into a truck, or possibly an automatic guided vehicle for transport to an indexed storage location.

Instead of your user program solving your logic and sending the solved logical status to the output status table and then to the output module when outputs are updated, the sequencer instruction sends a predetermined 16-bit data word representing the desired output configuration for a specific output module at a sequence step in a timed relationship to the output status file. The output status file turns the outputs on and off in the proper sequence during the output update portion of the scan. The SLC 500 sequencer output instruction, SQO, steps through the sequencer file, usually a bit file pattern whose bits have been previously set up to control the desired output devices.

In this exercise, we will develop a sequencer data table in a bit file where each 16-bit word in the data file will represent one step in our sequence. We will use a timer to step from one position (or "step") in the sequence to the next, to control (or "sequence") our outputs on and off.

DEVELOPING A SEQUENCER LADDER PROGRAM

We will develop the ladder rung in Figure 18-1 containing an input and the SQO instruction.

This application of the sequencer instruction will be programmed to step through a sequence of 16 steps. Each sequencer step will turn on one module LED at a time, starting with output LED 0 and stepping through LED 15. The output module in slot 2 will be the destination. When the sequence is completed, it will start over. Refer to your textbook, Figure 18-4, for an overview of the sequencer output instruction.

> NOTE: If you have a MicroLogix 1000, modify the exercise to cycle the available outputs on and off in sequence.

Program the following instruction parameters:

Select the SQO instruction.

Sequencer data table file B11. Start at element 0.

Mask FFFFh.

Destination, the output module in slot 2.

Sequencer control register R6:1.

Length of sequence is 16 steps.

Position is 0.

LAB ONE

1. _____ Open the Begin project.
2. _____ Create bit file B11 with 25 elements.
3. _____ Program the sequencer rung as illustrated in Figure 18-1. You will find the sequencer instructions under the File Shift/Sequencer tab of the tabbed Instruction Toolbar or in the Instruction Palette. Remember from your text, there are three sequencer instructions; use the SQO instruction.

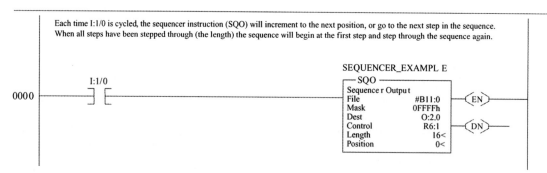

Figure 18-1 Sequencer ladder rung.

Information for each sequencer step regarding which output's points are affected must be created and stored in memory for the SQO instruction to refer to as it steps through the sequence. For this application, file B11 has been selected.

4. _____ Create data in the sequencer's data table as illustrated in Figure 18-2. Add a symbol for each step as you enter the data. Don't forget to leave B11:0 open for the start-up position.

> NOTE: You can key in each bit in the 16-bit word or convert each word to hex. Changing the data table's radix to hex will allow you to key in four characters per bit file element rather than 16 ones or zeros.

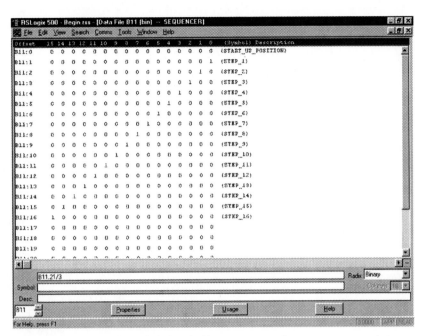

Figure 18-2 Data File B11 elements 0 through 16 programmed sequencer step data.

The sequencer instruction does not have a data file specifically for it, like counters and timers do. Timers store their status information in data file T4, while counters store their status information in data file C5. The control file R6 is a catch-all type file that will be used to store status data for many instructions that do not have a specifically defined data file for storage of their status information. Figure 18-3 shows a screen print of Data File R6, the control file.

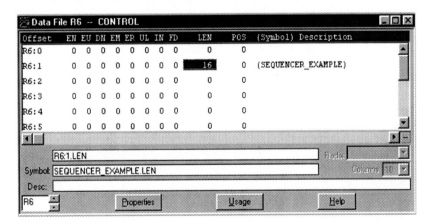

Figure 18-3 Screen print of Data File R6, the control file with 16 sequence steps filled in.

A control file element contains three words similar to a timer or counter. Word 0 contains status bits, word 1 is the length parameter (the number of steps in the sequence), and word 2 is the current position in the sequence.

There are eight status bits in the control file. Not all status bits are used with each instruction. The SQO instruction uses three of the available status bits.

5. _____ What are they? _____

6. _____ If you do not know, where will you find this information? _____

_____

7. _____ List the status bits used by the SQO instruction and define their functions. _____

_____

_____

_____

_____

8. _____ When completed programming, download and save the project as Sequence.
9. _____ Put processor in run mode and go on-line with the project.
10. _____ Press input switch I:1/0. Each time the button is pressed, the sequence should step to the next position. Observe the LEDs on the output module; the sequence should step from output 0 sequentially through 15 and begin over again.
11. _____ Stepping through the sequence, open the R6 file. Refer to Figure 18-4 and identify the correlation of the control file to the SQO instruction. Each identified piece is listed below.

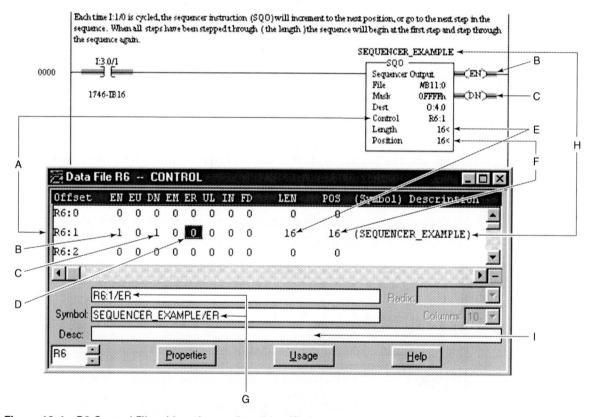

Figure 18-4 R6 Control File with major sections identified.

    A. Control file address
    B. SQO enable status bit
    C. SQO done bit
    D. SQO error bit
    E. Length parameter in the SQO instruction as well as the R6 file
    F. Position parameter in the SQO instruction and control file R6
    G. Address for selected data. Notice the ER bit is highlighted. Symbol is listed below the address.
    H. Instruction symbol
    I. Enter a text description here if desired.

12. _____ Ask your instructor if your exercise needs to be printed and handed in, or checked off.

LAB TWO

Modify your sequencer program as outlined below:

1. _____ A timer will increment the sequencer every 15 seconds.
2. _____ Modify the mask so for the first 10 cycles, only output points 0 through 7 are active. For cycles 11 through 25, outputs 8 through 15 are active. The mask will change automatically.
3. _____ Develop the sequencer data table on and off states to your liking.
4. _____ After 25 cycles, stop the sequencer.
5. _____ With the sequencer stopped, return the sequencer to the start-up position.
6. _____ Ask your instructor if your exercise needs to be printed and handed in, or checked off.

REVIEW QUESTIONS

1. How many sequencer instructions are available for SLC 500 processors? _____
2. List each sequencer instruction, its mnemonic, and a brief explanation of where each would be used. _____

_____

_____

_____

_____

3. Explain why the file indicator (#) is necessary when programming the sequencer instructions. _____

_____

_____

4. When entering the mask, can the mask be entered in a format other than hex? _____
5. If the mask can be entered in a format other than hex, list all possible formats and provide an example of each. _____

_____

_____

_____

6. What is the maximum number of words or steps for an SLC 500 modular processor? _____
7. What is the maximum number of words or steps for a MicroLogix controller? _____
8. Define the start-up position. _____

_____

9. Explain how the control file is used with the SQO instruction. _____

_____

10. How many status bits are used with the SQO instruction? _____

11. List the status bits used with the SQO instruction and define each one's function. _____

_____

_____

_____

_____

12. How many words make up a control file element? _____

13. Define element. _____

14. Using an RES instruction in conjunction with an SQO instruction accomplishes what? _____

_____

15. Explain the difference between programming the length parameter using RSLogix software in relation to programming using one of the DOS-based packages. _____

_____

_____

16. Explain why a mask is used with the SQO instruction. _____

_____

17. What is the purpose of the control parameter? _____

_____

18. Typically, the control address begins as R6. Explain the significance of the R and the 6. _____

_____

_____

19. Explain the file portion of the SQO instruction. _____

_____

_____

20. What mask would you use to control only the lower eight outputs on the output module in slot 3? _____

21. What mask would you use to control only the upper eight outputs on the output module in slot 3? _____

22. Explain how the sequencer instruction controls sixteen outputs on an output module when there is no output coil associated with the instruction. _____

_____

23. Explain the significance of the *length* portion of the SQO instruction. _____

_____

24. What does the value in the *position* portion of the SQO instruction tell you? _____

_____

25. Illustrate the control word structure. Identify all parts and their significance.

26. Assume you have an output module in slot 2 of your SLC 500 chassis. You need a 12-step sequencer. Inputs O:2/0 and O:2/1 are currently used to control field devices not associated with the sequence operation. Since this is the only output module with output points available, can you use this module for the sequence and still use O:2/0 and O:2/1 as separate

    outputs? _____

    If you answered yes to question 26, explain how. _____
    _____

LAB EXERCISE

19

Program Flow

LAB 19: PROGRAM FLOW

In this exercise, we will interpret program flow instructions on ladder logic for the SLC 500 family of PLCs and for RSLogix 500 software. The figures are actual screen dumps from RSLogix 500 software; they allow you to see what the rungs look like in the actual program on a computer. In this hands-on lab exercise, you will program a number of flow instructions using RSLogix 500 software.

The Following Questions Refer to Figure 19-1

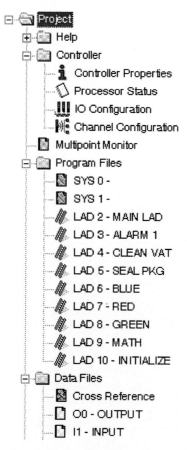

Figure 19-1 SLC 500 Project Tree showing program files.

1. When using an SLC 500 ladder file, ———— is typically the main ladder file.
2. Ladder files ———— through ———— are available for subroutines.
3. There are two reasons why one would use subroutines. Explain what they are.
4. Identify the parts of the scan from Figure 19-2.

   A.

   B.

   C.

   D.

   E.

   F.

   G.

   H.

   I.

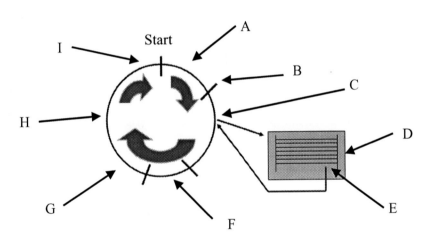

Figure 19-2   PLC scan when jumping to a subroutine.

The Following Questions Ask You to Interpret the Rung Operation of Figure 19-3

5A. Explain the function of the First Pass bit.
5B. When does the JSR instruction on rung 0 execute?
5C. What ladder file is executed when rung 0 is true?
5D. Why would we put the Initialize rungs into a subroutine?
5E. What ladder file are we looking at in Figure 19-3?
5F. If rung 3 were false, describe what the rung execution sequence would be.
5G. If rung 3 were true, describe what the rung execution sequence would be.
5H. Draw a picture illustrating the processor scan and how it would be affected by rung 3 being true.
5I. If rung 5 were false, describe what the rung execution sequence would be.
5J. If rung 5 were true, describe what the rung execution sequence would be.
5K. What do you have to take into consideration with regard to the Watchdog Timer when jumping into subroutines?

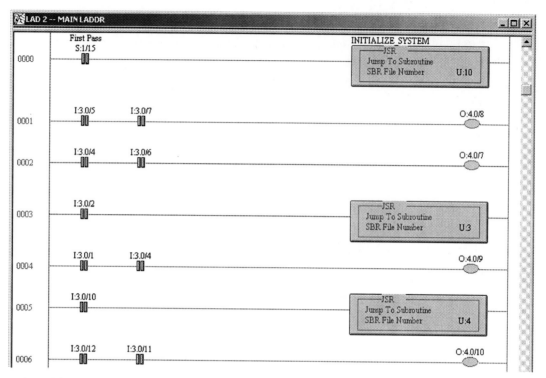

Figure 19-3   RSLogix 500 Jump To Subroutine Instructions.

The Following Questions Refer to Figure 19-4

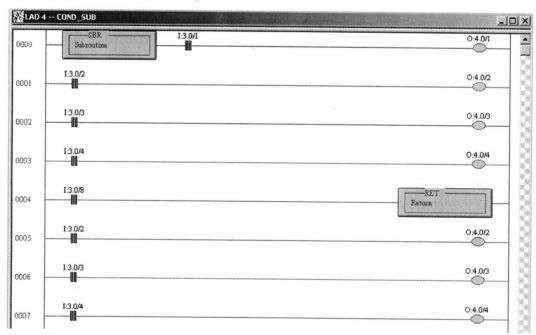

Figure 19-4   RSLogix 500 conditional subroutine.

6A. What is the function of the SBR Instruction?

6B. Is the SBR Instruction necessary? Explain your answer.

6C. What other instruction might be used as the first instruction in a subroutine file for an SLC 500?

6D. When using a PLC 5 or ControlLogix, how does the SBR Instruction change?

6E. If you use the SBR Instruction on the first rung of a subroutine, what must you remember with regard to that rung?

6F. Explain the function of rung 4 in Figure 19-4.

6G. What ladder file is illustrated in Figure 19-4?

7. Explain the function of rung 4 in Figure 19-5. Is this rung required? Explain your answer.

8. What ladder file is illustrated in Figure 19-5? Explain how we came to execute the ladder rungs.

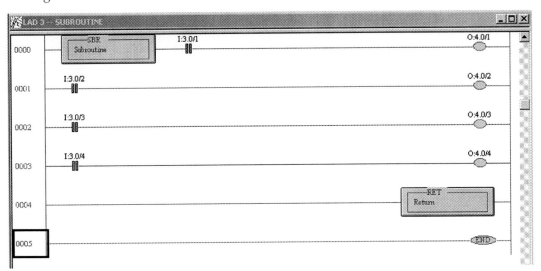

Figure 19-5   RSLogix 500 subroutine file.

The Following Questions Refer to Figure 19-6

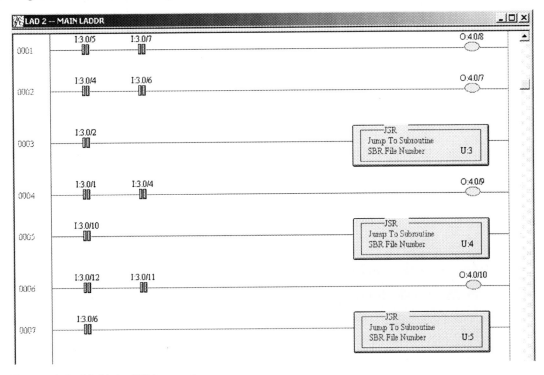

Figure 19-6   Multiple JSR Instructions.

9. When rung 3 is true, what happens?

10. When rung 5 is true, what happens?

11. Assume rung 7 is true. Which ladder file do we jump to?

The Following Questions Refer to Figure 19-7

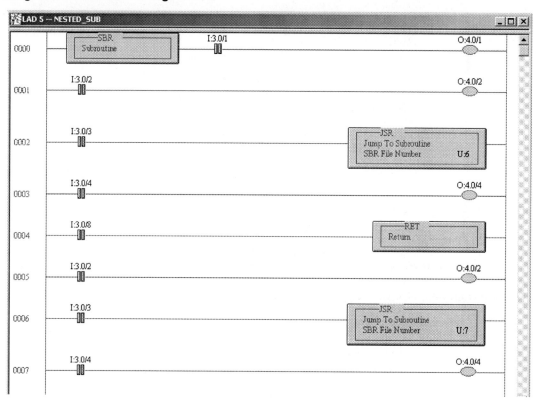

Figure 19-7 Nested subroutine.

12. What ladder file is represented in Figure 19-7?

13. Refer again to Figure 19-6. Explain how we came to execute the ladder rungs shown in Figure 19-7.

14. Jumping from one ladder file to another is called _____.

The Following Questions Refer to Figure 19-8

15. What ladder file is represented in Figure 19-8?

16. What is the name of this ladder file?

17. Refer again to Figure 19-7. Why did we jump into this ladder file?

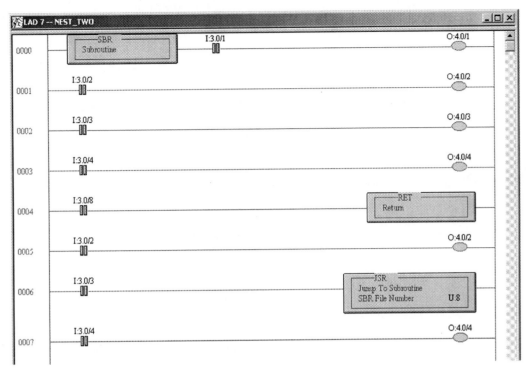

Figure 19-8 RSLogix 500 nesting subroutines.

18. What would happen if I:3.0/8 were true?

19. What would happen if both I:3.0/8 and I:3.0/3 were true? Explain what you would do.

The Following Questions Refer to Figure 19-9

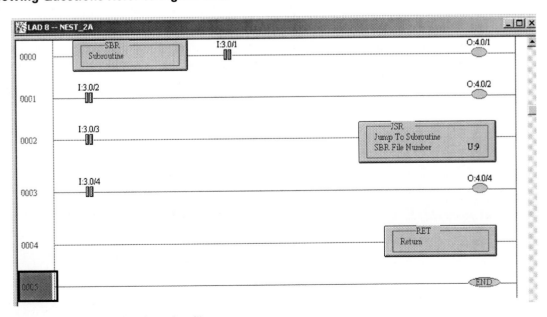

Figure 19-9 Nest_2A subroutine file.

20. Refer again to Figure 19-8. What caused the rungs in Figure 19-9 to be executed?

21. How many levels of nesting are allowed with an SLC 500 or a PLC 5?

22. Explain why rung 0004 is present. Does it need to be there?

The Following Questions Refer to Figure 19-10

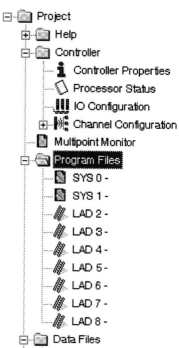

Figure 19-10 RSLogix 500 Program Files view.

23. Using the image shown in Figure 19-10, name the ladder files we have been using in the previous questions. This will help you to understand how the program files will look when viewing the RSLogix software. Since we did not use all the ladder files in this exercise, some will be blank.

The Following Questions Refer to Figure 19-11

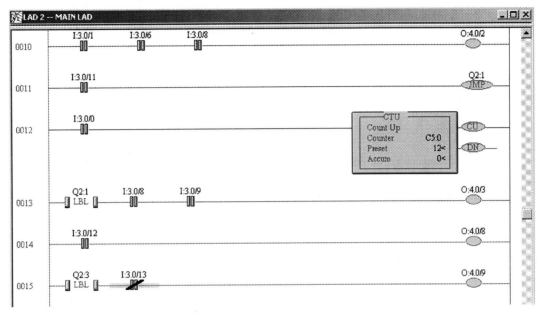

Figure 19-11 Jump to Label logic.

24. Explain what the JMP Instruction does.

25. What does Q2:1 signify?

26. Explain the function of the LBL Instruction.

27. Which LBL Instruction refers to the JMP on rung 11?

    How did you arrive at your answer?

28. If rung 11 were true, explain what would happen to the rungs in the jump zone.

    How would the counter behave?
    Which rung(s) would be in the jump zone?
    Explain how rung 14 logic would execute.

29. If rung 11 were false, explain what would happen to the rungs in the jump zone.

    Which rung(s) would be in the jump zone?
    How would the counter behave?
    Explain how rung 14 logic would execute.

30. Let's assume the following:

    Currently rung 13 is false.
    On the next scan, the Jump Instruction becomes true.
    Now input I:3.0/12 becomes true.
    What will happen to O:4.0/8? Explain your answer.

31. Assume the following:

   Currently rung 13 is true.
   On the next scan, the Jump Instruction becomes false.
   Now input I:3.0/12 becomes false.
   What will happen to O:4.0/8? Explain your answer.

The Following Questions Refer to Figure 19-12

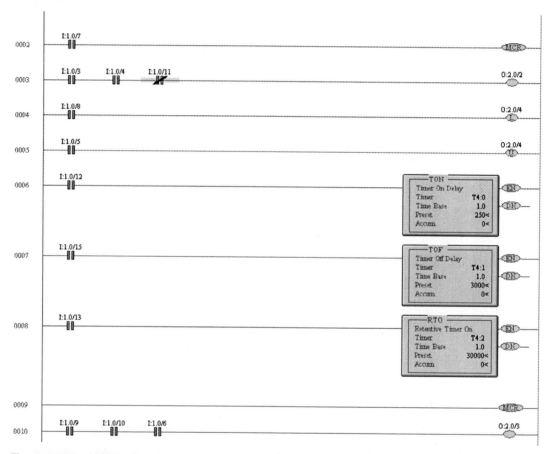

Figure 19-12   MCR logic.

32. Explain how an MCR zone works.

33. Refer to Figure 19-12. Which rungs are within the MCR zone?

34. What will happen to the TON Instruction if Input I:1.0/7 is false?

35. What will happen to the TOF Instruction if Input I:1.0/7 is false?

36. Assume there are 50 rungs in this ladder file. What happens if the MCR Instruction on rung 0009 is omitted?

The Following Questions Refer to Figure 19-13

Figure 19-13   SLC 500 IOM Instruction.

37. Explain the function of the IOM Instruction.

38. If we wanted to update slot 7, what would the slot parameter be?

39. We want to control I/O point 12. What would you program into the Mask?

40. What is the length parameter used for?

The Following Questions Refer to Figure 19-14

Figure 19-14   SLC 500 IIM Instruction.

41. Explain the function of the IIM Instruction.

42. If we wanted to update slot 21, what would the slot parameter be?

43. We want to control I/O point 9. What would you program into the Mask?

44. Look at the three rungs shown in Figure 19-15. Then Interpret the IIM and IOM Instructions and determine what addresses must be programmed on the center rung.

    Inputs =

    Outputs =

45. An IIM Instruction slot parameter is I:3.0. For each mask listed below, determine the address to be updated.

    A.   0040h

    B.   A020h

    C.   0001h

    D.   0080h

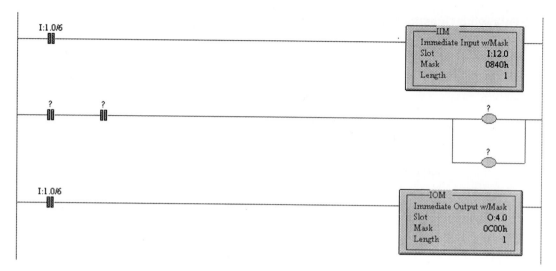

Figure 19-15 IIM and IOM Instruction pair.

Questions 46 and 47 Refer to Figure 19-16

Figure 19-16 SLC TEF Instruction.

46. Describe in detail what an REF Instruction will do when executed. Assume we are working with an SLC 5/05 processor.

47. Describe the sequence of events that occurs when an REF Instruction is executed.

    1.

    2.

    3.

    4.

    5.

48.

49. Channel 1 on a 5/04 processor is _____.

50. Channel 1 on a 5/05 processor is _____.

PROGRAMMING EXERCISE

For this exercise, we are going to create a new project and new subroutine files.

THE LAB

1. _____ Open your RSLogix 500 software.
2. _____ If you are using a MicroLogix 1000 PLC, go on to the next question. If you are using an SLC 500 modular or MicroLogix 1200 or 1500 processor, jump to step 5.
3. _____ Create a new project. Select the appropriate MicroLogix 1000.

New MicroLogix 1000 project default ladder files are shown in Figure 19-17. Notice that the MicroLogix 1000 only has 16 ladder files. Currently, ladder file 16 is a debug file (note the little bug icon). Debug files are used with a software emulator. This file could be changed to a ladder file by right-clicking on it, going to Properties, and unchecking the Debug box.

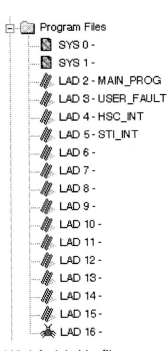

Figure 19-17   MicroLogix 1000 default ladder files.

4. _____ Rename the ladder files in Figure 19-17 by right-clicking on the current file and selecting Rename. Rename:
   A. ladder file 2 as Main-Prog
   B. ladder file 3 as Initialize
   C. ladder file 4 as Conveyor
   D. ladder file 5 as Ld Recipe
   E. ladder file 6 as Mix
   F. ladder file 7 as Fill
   G. ladder file 8 as Cap
   H. ladder file 9 as Label
   I. ladder file 10 as Alarms
   J. ladder file 11 as Clean Vat
   K. ladder file 12 as Package

Figure 19-18 shows your renamed ladder files.

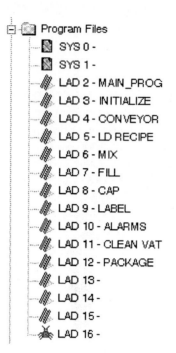

Program Files
- SYS 0 -
- SYS 1 -
- LAD 2 - MAIN_PROG
- LAD 3 - INITIALIZE
- LAD 4 - CONVEYOR
- LAD 5 - LD RECIPE
- LAD 6 - MIX
- LAD 7 - FILL
- LAD 8 - CAP
- LAD 9 - LABEL
- LAD 10 - ALARMS
- LAD 11 - CLEAN VAT
- LAD 12 - PACKAGE
- LAD 13 -
- LAD 14 -
- LAD 15 -
- LAD 16 -

Figure 19-18   Completed subroutine ladder files for MicroLogix 1000.

This completes the MicroLogix 1000 portion of this Lab.

Lab Exercise for SLC 500 Modular Processor

5. _____ Create a new project for your specific SLC 500 processor.
6. _____ Create the following:
   A. Ladder file 2 as the Main Lad
   B. Ladder file 3 as Initialize
   C. Ladder file 4 as Conveyor
   D. Ladder file 5 as Ld Recipe
   E. Ladder file 6 as Mix
   F. Ladder file 7 as Fill
   G. Ladder file 8 as Cap
   H. Ladder file 9 as Label
   I. Ladder file 10 as Alarms
   J. Ladder file 11 as Clean Vat
   K. Ladder file 12 as Package
   L. Ladder file 13 as Palletizing
7. _____ To create new ladder files, right-click on Program Files and select New. See Figure 19-19.
      The next available ladder file that can be created will be displayed. You can select a different file number if you wish. Refer to Figure 19-20.
8. _____ Type the name of the file you wish to create into the Name field. The first subroutine file you will create is called Initialize. A description of the file can be added if desired.
9. _____ Click OK when completed.

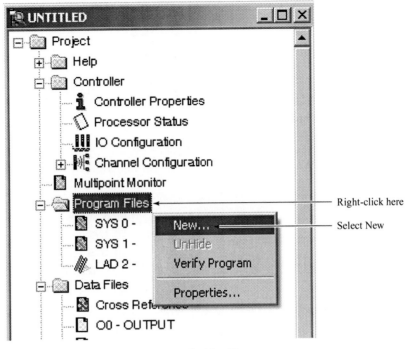

Figure 19-19   Select New to create new ladder file.

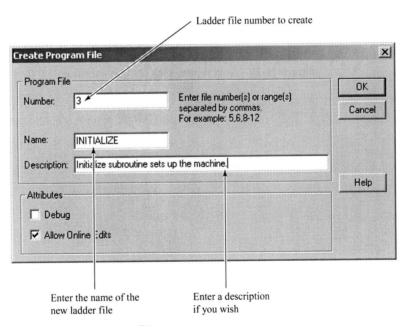

Figure 19-20   Create Program File.

10. _____ Your program files should look like those shown in Figure 19-21.

11. _____ Continue creating the remaining ladder files using this procedure.

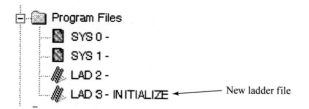

Figure 19-21   Initialize subroutine file.

12. _____ When completed, your program files should look like those shown in Figure 19-22.

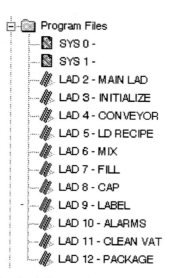

Figure 19-22   New ladder files for Lab exercise.